# A Practical Guide to the Rules of the Road

## For OOW, Chief Mate and Master Students

Third edition

Farhan Saeed

Routledge
Taylor & Francis Group
LONDON AND NEW YORK

First published 2010 by Seamanship Tutor Ltd

Third edition published 2015
by Routledge
2 Park Square, Milton Park, Abingdon, Oxon OX14 4RN

and by Routledge
605 Third Avenue, New York, NY 10017

*Routledge is an imprint of the Taylor and Francis Group, an informa business*

First issued in hardback 2015



Although great care has been taken with the writing and production of this publication, neither the publisher nor the author can accept any responsibility for any errors, omissions or their consequences.

This publication has been prepared to deal with the subject of collision regulations. This should not, however, be taken to mean that this publication deals comprehensively with all the issues of the subject. The opinions expressed are those of the author only and are not necessarily to be taken as policies or views of any organisation with which he has any connection.



*British Library Cataloguing in Publication Data*
A catalogue record for this book is available from the British Library

*Library of Congress Cataloging in Publication Data*
Saeed, Farhan.
A practical guide to the rules of the road for OOW, chief mate and master students / Farhan Saeed. – 3rd edition.
pages cm
First published: 2010.
1. Rule of the road at sea. 2. Navigation–Safety measures. 3. Signals and signaling. I. Title. II. Title: A practical guide to the rules of the road for Officer Of the Watch, chief mate and master students.
VK371.S25 2015
623.88'84–dc23
2014037612

ISBN 978-1-138-13717-2 (hbk)
ISBN 978-1-138-84389-9 (pbk)
ISBN 978-1-315-73078-3 (ebk)

Typeset in Sabon by
Servis Filmsetting Ltd, Stockport, Cheshire

# Contents

# Preface

The purpose of this book is to provide a form of self-teaching of the Rules of the Road and self-examination on the navigation lights and related situations that may occur at sea or be asked about in the oral examinations for the Certificate of Competency.

This book is divided into two main sections. The first section (Parts A–E) contains simple explanations of the rules, while the three chapters of the second section cover more than fifty situations, exam questions and model answers. The rules, explanations and model answers are the views of the author and not necessarily the view of courts and regulators.

Interactive navigation light identification software is available online and the web link is provided after the explanation of the rules. This contains enough navigation light examples as are required for the Certificate of Competency examinations. All questions are asked randomly. You will need to identify the light and then check the answer at each page. You need to master these simple tests before you go for your oral examination.

The book is aimed at all grades of Certificate of Competency, and if you are preparing for the OOW or Chief Mate examination then, in some situations (where there is doubt, such as when sounding at least five or more short and rapid blasts), you will need to call/inform the master as appropriate.

# Acknowledgements

The author and publishers would like to thank the following for reviewing the book and providing valuable advice during the several stages of the book's preparation:

Dr Alan Wall, Dr Paul Hughes, Mr Vikas Patra and Capt. Shan Hashmi.

# Introduction

The International Regulations for Preventing Collision at Sea 1972 (including amendments up to 2009) are divided into five main parts and four auxiliary annexes. This is as follows:

| | |
|---|---|
| Part A | General |
| Part B | Steering and sailing rules<br>Section I: Conduct of vessels in any conditions of visibility<br>Section II: Conduct of vessels in sight of one another<br>Section III: Conduct of vessels in restricted visibility |
| Part C | Lights and shapes |
| Part D | Sound and light signals |
| Part E | Exemptions |
| Annex I | Positioning and technical details of lights and shapes |
| Annex II | Additional signals for fishing vessels fishing in close proximity |
| Annex III | Technical details of sound signal appliances |
| Annex IV | Distress signals |

The regulations consist of 38 rules in total.

Part A covers general rules (1–3) covering application of the rules, responsibilities and definitions.

Part B contains the steering and sailing rules (4–19), subdivided into three sections according to the degree of visibility. It is very important to have a thorough understanding of this part. The rules of this part set out the actions of every mariner in all conditions of visibility. The rules of Section I (4–10) of Part B apply in any condition of visibility. Rules of Section II (11–18) are applicable to vessels in sight of one another.

The third section consists of Rule 19, which is applicable in restricted visibility only.

Part B is of the utmost importance to the watchkeeper. The rules of Section I are applicable in all conditions of visibility. Section II and Section III can only be applied separately, as one is applicable in clear visibility and the other in restricted visibility, but both will be applied in conjunction with Section I.

Part C (Rules 20–31) contains details about the lights and shapes to be displayed by different types of vessel. It is very necessary for the watchkeeping officer to identify different types of vessels by day or night by means of shapes or lights displayed. Only by identifying the type of vessel navigating will the watchkeeper be able to take appropriate action to avoid collision.

Part D gives details of manoeuvring and warning signals made by sound and/or light in both conditions of visibility – clear and restricted.

Part E concerns exemptions to the positioning of the lights and requirements of sound signals for vessels whose keel is laid during the entry into force of these regulations, provided that she complies with the requirements of the International Regulations for Preventing Collision at Sea, 1960.

# Part A
# General

## Rule 1

### *Application*

(a) **These rules shall apply to all vessels upon the high seas and in all waters connected therewith navigable by sea-going vessels.**

All vessels must comply with these rules; that is, when they are navigating upon high seas and navigable waterways connected with high seas.

(b) **Nothing in these rules shall interfere with the operation of special rules made by an appropriate authority for roadsteads, harbours, rivers, lakes or inland waterways connected with the high seas and navigable by seagoing vessels. Such special rules shall conform as closely as possible to these rules.**

National authorities can implement special rules for the harbours, rivers or inland waterways connected with high seas and used by seagoing vessels. However, such rules must be as similar as possible to these rules.

(c) **Nothing in these rules shall interfere with the operation of any special rules made by the government of any state with respect to additional station or signal lights, shapes or whistle signals for ships of war and vessels proceeding under convoy, or with respect to additional station or signal lights or shapes for fishing vessels engaged in fishing as a fleet. These additional station or signal lights, shapes or whistle signals shall, so far as possible, be such that they cannot be mistaken for any light, shape or signal authorised elsewhere under these rules.**

*A Practical Guide to the Rules of the Road, Third edition.* 978-1-138-84389-9.

The government of any state can introduce any special rules regarding additional station or signal lights, shapes or whistle signals for ships of war, vessels proceeding under convoy or vessels engaged in fishing, provided these additional lights, shapes or signals cannot be mistaken for any light, shape or signal authorised anywhere under these rules.

(d) **Traffic separation schemes may be adopted by the Organization for the purpose of these rules.**

The Organization, IMO, may adopt traffic separation schemes for the purpose of these rules.

(e) **Whenever the government concerned shall have determined that a vessel of any special construction or purpose cannot comply with the provisions of any of these rules with respect to the number, position, range or arc of visibility of lights or shapes, as well as to the disposition and characteristics of sound-signalling appliances, such a vessel shall comply with such other provisions in regard to the number, position, range or arc of visibility of lights or shapes, as well as to the disposition and characteristics of sound-signalling appliances, as her government shall have determined to be the closest possible compliance with these rules in respect of that vessel.**

Where a vessel of special construction cannot comply with these rules with respect to the number, position, range or arc of visibility of lights and shapes, as well as to the disposition and characteristics of sound-signalling appliances, such a vessel must comply with other provisions as determined by her government, to be as close as possible to these rules.

## Rule 2

### *Responsibility*

(a) **Nothing in these rules shall exonerate any vessel, or the owner, master or crew thereof, from the consequences of any neglect to comply with these rules or of the neglect**

**of any precaution which may be required by the ordinary practice of seamen, or by the special circumstances of the case.**

These rules do not relieve any vessel, owner, master or crew of/from the obligation of complying with these rules or of/from neglecting any precaution required by the ordinary practice of seamen.

(b) **In construing and complying with these rules due regard shall be had to all dangers of navigation and collision and to any special circumstances, including the limitations of the vessels involved, which may make a departure from these rules necessary to avoid immediate danger.**

To avoid collision, there may be occasions when it becomes ultimately essential to deviate from these rules. Any action taken in accordance of Rule 2(b) must be effective and result in passing other vessels safely.

## Rule 3

### *General definitions*

**For the purpose of these rules, except where the context otherwise requires:**

(a) **The word 'vessel' includes every description of water craft, including non-displacement craft, WIG craft and seaplanes, used or capable of being used as a means of transportation on water.**
(b) **The term 'power-driven vessel' means any vessel propelled by machinery.**
(c) **The term 'sailing vessel' means any vessel under sail provided that propelling machinery, if fitted, is not being used.**
(d) **The term 'vessel engaged in fishing' means any vessel fishing with nets, lines, trawls or other fishing apparatus which restrict manoeuvrability, but does not include a**

vessel fishing with trolling lines or other fishing apparatus which do not restrict manoeuvrability.

(e) The word 'seaplane' includes any aircraft designed to manoeuvre on the water.

(f) The term 'vessel not under command' means a vessel which through some exceptional circumstance is unable to manoeuvre as required by these rules and is therefore unable to keep out of the way of another vessel.

(g) The term 'vessel restricted in her ability to manoeuvre' means a vessel which from the nature of her work is restricted in her ability to manoeuvre as required by these rules and is therefore unable to keep out of the way of another vessel. The term 'vessels restricted in their ability to manoeuvre' shall include but not be limited to:

  (i) a vessel engaged in laying, servicing or picking up a navigation mark, submarine cable or pipeline;
  (ii) a vessel engaged in dredging, surveying or underwater operations;
  (iii) a vessel engaged in replenishment or transferring persons, provisions or cargo while underway;
  (iv) a vessel engaged in the launching or recovery of aircraft;
  (v) a vessel engaged in mine clearance operations;
  (vi) a vessel engaged in a towing operation such as severely restricts the towing vessel and her tow in their ability to deviate from their course.

(h) The term 'vessel constrained by her draught' means a power-driven vessel which, because of her draught in relation to the available depth and width of navigable water, is severely restricted in her ability to deviate from the course she is following.

(i) The word 'underway' means that a vessel is not at anchor, or made fast to the shore, or aground.

(j) The words 'length' and 'breadth' of a vessel mean her length overall and greatest breadth.

(k) Vessels shall be deemed to be in sight of one another only when one can be observed visually from the other.

(l) The term 'restricted visibility' means any condition in which visibility is restricted by fog, mist, falling snow, heavy rainstorms, sandstorms or any other similar causes.

(m) The term 'Wing-In-Ground (WIG) craft' means a multi-modal craft which, in its main operational mode, flies in close proximity to the surface by utilizing surface-effect action.

All these definitions in rule 3 are self-explanatory and you must learn these by heart before offering yourself up for an oral examination.

# Part B
# Steering and sailing rules

## SECTION I: CONDUCT OF VESSELS IN ANY CONDITION OF VISIBILITY

### Rule 4

#### *Application*

Rules in this section apply in any condition of visibility.

### Rule 5

#### *Lookout*

Every vessel shall at all times maintain a proper lookout by sight and hearing as well as by all available means appropriate in the prevailing circumstances and conditions so as to make a full appraisal of the situation and of the risk of collision.

It is absolutely essential to maintain an all round lookout, at all times, by sight and hearing and by all available means including radar and AIS in all kinds of visibility. Extra lookouts should be posted in restricted visibility. The early detection of a target will help to make an early judgement of the situation, and of the risk of collision.

### Rule 6

#### *Safe speed*

Every vessel shall at all times proceed at a safe speed so that she can take proper and effective action to avoid collision and be stopped within a distance appropriate to the prevailing circumstances and conditions.

*A Practical Guide to the Rules of the Road, Third edition.* 978-1-138-84389-9.

This rule directs every vessel to proceed at a *safe speed*. It is hard to say what a safe speed is in terms of a number of knots. However, the rule suggests that a vessel shall proceed at such a speed that she can take effective action to avoid collision and, that the vessel can be stopped within an appropriate distance. Every vessel must make its own judgement of what a safe speed is, by considering the factors given below.

**In determining a safe speed the following factors shall be among those taken into account:**

- **(a) By all vessels:**
  - **(i) the state of visibility;**
  - **(ii) the traffic density including concentrations of fishing vessels or any other vessels;**
  - **(iii) the manoeuvrability of the vessel with special reference to stopping distance and turning ability in the prevailing conditions;**
  - **(iv) at night the presence of background light such as from shore lights or from back scatter of her own lights;**
  - **(v) the state of wind, sea and current, and the proximity of navigational hazards;**
  - **(vi) the draught in relation to the available depth of water.**

The above factors are to be considered by all vessels. State of visibility is one of the main factors to determine the safe speed of one's own vessel. Prudent masters will reduce speed to minimum in restricted visibility.

- **(b) Additionally, by vessels with operational radar:**
  - **(i) the characteristics, efficiency and limitations of the radar equipment;**
  - **(ii) any constraints imposed by the radar range scale in use;**
  - **(iii) the effect on radar detection of the sea state, weather and other sources of interference;**

(iv) **the possibility that small vessels, ice and other floating objects may not be detected by radar at an adequate range;**

(v) **the number, location and movement of vessels detected by radar;**

(vi) **the more exact assessment of the visibility that may be possible when radar is used to determine the range of vessels or other objects in the vicinity.**

In short, operational radar does not automatically mean that you can go full speed in all weathers and visibility.

## Rule 7

### *Risk of collision*

(a) **Every vessel shall use all available means appropriate to the prevailing circumstances and conditions to determine if risk of collision exists. If there is any doubt such risk shall be deemed to exist.**

First of all we have to ascertain how to determine risk of collision and what are the available means to determine the collision. This rule describes in section (d) that *'such risk shall be deemed to exist if the compass bearing of an approaching vessel does not appreciably change'*. A bearing of an object can be taken with the magnetic compass on the monkey island, a gyro repeater in the bridge wings, a gyro repeater in the bridge, and the radar. Here the rule suggests that a vessel shall use all available means to determine collision risk.

(b) **Proper use shall be made of radar equipment if fitted and operational, including long-range scanning to obtain early warning of risk of collision and radar plotting or equivalent systematic observation of detected objects.**

Radar must be used efficiently, if fitted and operational. Regular switching to long ranges gives an early warning of the risk of collision, through manually plotting targets or by use of Automatic Plotting Aids.

(c) **Assumptions shall not be made on the basis of scanty information, especially scanty radar information.**

Scanty information means inadequate information. A watch officer must not make an assumption based on barely sufficient information. A watch officer must complete a proper radar plot of the target before making any judgement about the risk of collision.

(d) **In determining if risk of collision exists the following considerations shall be among those taken into account:**

(i) **such risk shall be deemed to exist if the compass bearing of an approaching vessel does not appreciably change;**

(ii) **such risk may sometimes exist even when an appreciable bearing change is evident, particularly when approaching a very large vessel or a tow or when approaching a vessel at close range.**

Whenever a target appears on the horizon, the watch officer must take a series of compass bearings to determine if a risk of collision exists. If the bearings of an approaching vessel, taken over sufficient time, do not change considerably then a risk of collision is deemed to exist. Sometimes, the bearing may change appreciably while the risk of collision still exists; for example, when approaching a large vessel at close range or when approaching a tow. While approaching a tow, make sure to take bearings of the forward end of the towing vessel and also of the astern of the vessel being towed.

(Please visit www.seamanshiptutor.com/rorbook to download animations on rule 7.)

## Rule 8

### *Action to avoid collision*

(a) **Any action taken to avoid collision shall be taken in accordance with the rules of this part and shall, if the circumstances of the case admit, be positive, made in**

**ample time and with due regard to the observance of good seamanship.**

Any action taken to avoid collision must comply with the rules of Part B (Sailing and Steering Rules). There are three key points stressed here: (i) positive action, (ii) action made in ample time, and (iii) action based on good seamanship.

**(b) Any alteration of course and/or speed to avoid collision shall, if the circumstances of the case admit, be large enough to be readily apparent to another vessel observing visually or by radar; a succession of small alterations of course and/or speed should be avoided.**

A succession of small alterations may not be readily apparent to another vessel, whereas a large alteration of course and/or speed should be apparent to the other vessel. In the case of small alterations, the other vessel will not be able to judge whether one's own vessel has taken action to avoid collision.

**(c) If there is sufficient sea-room, alteration of course alone may be the most effective action to avoid a closequarters situation, provided that it is made in good time, is substantial and does not result in another closequarters situation.**

If circumstances of the case admit, an alteration of course is the preferable action in order to avoid collision, provided it is made in ample time.

**(d) Action taken to avoid collision with another vessel shall be such as to result in passing at a safe distance. The effectiveness of the action shall be carefully checked until the other vessel is finally past and clear.**

The action taken to avoid collision must be positive and result in passing at a safe distance with the other vessel. The watch officer must observe the effectiveness of the action until the other vessel is definitely past and clear.

**(e) If necessary to avoid collision or allow more time to assess the situation, a vessel shall slacken her speed or**

**take all way off by stopping or reversing her means of propulsion.**

If a vessel needs more time to appraise the situation and it is determined that the other vessel is quite close, then the options of reducing speed or stopping engines must be used. This will buy her time and she can make the exact assessment of the situation.

(f) (i) **A vessel which, by any of these rules, is required not to impede the passage or safe passage of another vessel shall, when required by the circumstances of the case, take early action to allow sufficient sea-room for the safe passage of the other vessel.**

(ii) **A vessel required not to impede the passage or safe passage of another vessel is not relieved of this obligation if approaching the other vessel so as to involve risk of collision and shall, when taking action, have full regard to the action which may be required by the rules of this part.**

(iii) **A vessel the passage of which is not to be impeded remains fully obliged to comply with the rules of this part when the two vessels are approaching one another so as to involve risk of collision.**

A power-driven vessel, for example, is required by these rules *to avoid impeding* the safe passage of a vessel constrained by her draft. This rule directs the power-driven vessel to take early action to allow sufficient sea-room to the vessel constrained by her draught. A power-driven vessel shall take action to avoid collision with a vessel constrained by her draft in accordance with these rules and shall not alter course to port for a vessel constrained by her draft which is on her own port side. By definition a vessel constrained by her draft is a power-driven vessel (ref. rule 3h) and in such a situation the vessel constrained by her draft remains a give way vessel and power-driven vessel is a stand on vessel but takes action not to impede the passage of the vessel constrained by her draft. Since a vessel constrained by her draft remains fully obliged to comply with these rules, she is only faced with the problem of available depth and width of water, but she can always reduce speed to avoid collision. So she needs to determine as early as possible that a power-driven vessel on her own starboard side is taking substantial action so that she can continue on her own course. If power-driven vessel does not take action to avoid collision when she has a vessel constrained by her draft on her own port side, then according to rule 8f(iii) the vessel constrained by her draft must take action to avoid collision by reducing her speed or stopping engines.

Further examples of vessels that are not to impede the passage of other vessels include:

### Narrow channels

- A vessel of less than 20 metres in length or a sailing vessel *shall not* impede the passage of a vessel which can safely navigate only within a narrow channel or fairway.
- A vessel engaged in fishing *shall not* impede the passage of any other vessel navigating within a narrow channel or fairway.
- A vessel *shall not* cross a narrow channel or fairway if such crossing impedes the passage of a vessel which can safely navigate only within such a channel or fairway.

### Traffic separation schemes

- A vessel engaged in fishing *shall not* impede the passage of any vessel following a traffic lane.
- A vessel of less than 20 metres in length or a sailing vessel *shall not* impede the safe passage of a power-driven vessel following a traffic lane.

Please note from above explanations that the difference between '*shall avoid impeding*' and '*shall not impede*'. A power-driven vessel *shall avoid impeding* the safe passage of a vessel constrained by her draft. In TSS and narrow channel, a sailing vessel, a vessel engaged in fishing and a vessel less than 20m in length *shall not impede* the passage of a vessel following a lane or within the channel. '*Shall not impede*' puts more emphasis on such vessels to stay clear.

## Rule 9

### *Narrow channels*

(a) **A vessel proceeding along the course of a narrow channel or fairway shall keep as near to the outer limit of the channel or fairway which lies on her starboard side as is safe and practicable.**

A vessel proceeding in a narrow channel shall keep to the outer limit of the channel or fairway which lies on its own starboard side. This way all vessels shall pass port to port.

(b) **A vessel of less than 20 metres in length or a sailing vessel shall not impede the passage of a vessel which can safely navigate only within a narrow channel or fairway.**

Rule 8(f) applies here.

(c) **A vessel engaged in fishing shall not impede the passage of any other vessel navigating within a narrow channel or fairway.**

Rule 8(f) applies here.

(d) **A vessel shall not cross a narrow channel or fairway if such crossing impedes the passage of a vessel which can safely navigate only within such a channel or fairway. The latter vessel may use the sound signal prescribed in Rule 34(d) if in doubt as to the intention of the crossing vessel.**

Rule 8(f) applies here, since crossing impedes the safe passage of a vessel which can only safely navigate within the narrow channel. The vessel navigating in the narrow channel may sound at least five short and rapid blasts on the whistle.

**(e) (i) In a narrow channel or fairway when overtaking can take place only if the vessel to be overtaken has to take action to permit safe passing, the vessel intending to overtake shall indicate her intention by sounding the appropriate signal prescribed in Rule 34(c)(i). The vessel to be overtaken shall, if in agreement, sound the appropriate signal prescribed in Rule 34(c)(ii) and take steps to permit safe passing. If in doubt she may sound the signals prescribed in Rule 34(d).**

If a vessel intends to overtake within a narrow channel she must seek permission of a vessel being overtaken by sounding the appropriate signal:

- Two prolonged blasted blasts followed by one short blast means 'I intend to overtake you on your starboard side.'
- Two prolonged blasted blasts followed by two short blasts means 'I intend to overtake you on your port side.'

If a vessel being overtaken agrees to the request then she shall sound one prolonged blasted, one short, one prolonged blasted and one short blast. If a vessel being overtaken is in any doubt or in disagreement then she shall sound at least five short and rapid blasts.

**(ii) This rule does not relieve the overtaking vessel of her obligation under Rule 13.**

Rule 13 remains applicable here, meaning it is still the responsibility of the overtaking vessel to keep clear.

**(f) A vessel nearing a bend or an area of a narrow channel or fairway where other vessels may be obscured by an intervening obstruction shall navigate with particular alertness and caution and shall sound the appropriate signal prescribed in Rule 34(e).**

A vessel nearing a bend shall sound one prolonged blasted blast to indicate her presence.

(g) **Any vessel shall, if the circumstances of the case admit, avoid anchoring in a narrow channel.**

## Rule 10

### *Traffic separation schemes*

(a) **This rule applies to traffic separation schemes adopted by the Organization and does not relieve any vessel of her obligation under any other rule.**

This rule is only applicable to traffic separation schemes adopted by IMO. All vessels following traffic separation schemes are not relieved of their obligations under any other rule; which means all other rules remain applicable in traffic separation schemes.

(b) **A vessel using a traffic separation scheme shall:**
- (i) **proceed in the appropriate traffic lane in the general direction of traffic flow for that lane;**
- (ii) **so far as practicable keep clear of a traffic separation line or separation zone;**
- (iii) **normally join or leave a traffic lane at the termination of the lane, but when joining or leaving from either side shall do so at as small an angle to the general direction of traffic flow as practicable.**

A vessel using a traffic separation scheme shall proceed in the appropriate lane in the general direction of the traffic flow for that lane (+/–5°) and if possible keep clear of the traffic separation line or the separation zone. Joining and leaving is preferred at the termination points but when joining or leaving from the sides, it shall be done at small angle to the general direction of traffic flow.

(c) **A vessel shall, so far as practicable, avoid crossing traffic lanes but if obliged to do so shall cross on a heading as nearly as practicable at right angles to the general direction of traffic flow.**

Crossing traffic lanes is not recommended here, but when crossing cannot be avoided it shall be performed with a heading at right angles to the general direction of traffic flow. Note here that it is the 'heading' which must be at right angles to the traffic flow when crossing. That means that it is not the 'true course made good' but the 'true course to steer' which matters, as demonstrated in the following animations.

(Please visit www.seamanshiptutor.com/rorbook to download animations on Rule 10.)

**(d) (i) A vessel shall not use an inshore traffic zone when she can safely use the appropriate traffic lane within the adjacent traffic separation scheme. However, vessels of less than 20 metres in length, sailing vessels and vessels engaged in fishing may use the inshore traffic zone.**

**(ii) Notwithstanding sub-paragraph (d)(i), a vessel may use an inshore traffic zone when en route to or from a port, offshore installation or structure, pilot station or any other place situated within the inshore traffic zone, or to avoid immediate danger.**

Through vessels shall use a traffic separation scheme and not use inshore traffic zones. But sailing vessels, vessels engaged in fishing and vessels less than 20 metres in length may use inshore traffic zones.

**(e) A vessel other than a crossing vessel or a vessel joining or leaving a lane shall not normally enter a separation zone or cross a separation line except:**

**(i) in cases of emergency to avoid immediate danger;**

**(ii) to engage in fishing within a separation zone.**

A vessel may enter a separation zone or cross a separation line only in the following conditions:

- A vessel in an emergency;
- To engage in fishing within a separation zone;
- Crossing a lane;
- Joining or leaving a lane.

(f) A vessel navigating in areas near the terminations of traffic separation schemes shall do so with particular caution.
(g) A vessel shall so far as practicable avoid anchoring in a traffic separation scheme or in areas near its terminations.
(h) A vessel not using a traffic separation scheme shall avoid it by as wide a margin as is practicable.
(i) A vessel engaged in fishing shall not impede the passage of any vessel following a traffic lane.

Rule 8(f) applies here.

(j) A vessel of less than 20 metres in length or a sailing vessel shall not impede the safe passage of a power-driven vessel following a traffic lane.

Rule 8(f) applies here.

Safe passage refers to power-driven vessel which is required to follow the appropriate lane if navigating in the area. For a power-driven vessel, a lane is the safe passage. On the other hand, passage is for any vessel which is following the lane, such as if a sailing vessel is following a lane then a fishing vessel shall not impede the passage of any vessel following a traffic lane (Rule 10(i)).

(k) A vessel restricted in her ability to manoeuvre when engaged in an operation for the maintenance of safety of navigation in a traffic separation scheme is exempted from complying with this rule to the extent necessary to carry out the operation.
(l) A vessel restricted in her ability to manoeuvre when engaged in an operation for the laying, servicing or picking up of a submarine cable, within a traffic separation scheme, is exempted from complying with this rule to the extent necessary to carry out the operation.

The above two types of vessel are exempted from complying with this rule because of the nature of their work. A vessel engaged in an operation for the maintenance of safety of navigation can steer in any direction within the traffic separation scheme (TSS) as may be required by the extent of the operation. Similarly, a vessel engaged in an operation for the laying, servicing or picking up of a submarine cable within a traffic separation scheme is exempted so as to be able to carry out the operation.

## SECTION II: CONDUCT OF VESSELS IN SIGHT OF ONE ANOTHER

### Rule 11

#### *Application*

**Rules in this section apply to vessels in sight of one another.**

This rule is here to emphasise the point that this section applies only to vessels when in sight of one another. Remember that Section I applies in all conditions of visibility, while Section II applies only to vessels in sight of one another.

### Rule 12

#### *Sailing vessels*

**(a) When two sailing vessels are approaching one another, so as to involve risk of collision, one of them shall keep out of the way of the other as follows:**

- **(i) when each has the wind on a different side, the vessel which has the wind on the port side shall keep out of the way of the other;**
- **(ii) when both have the wind on the same side, the vessel which is to windward shall keep out of the way of the vessel which is to leeward;**
- **(iii) if a vessel with the wind on the port side sees a vessel to windward and cannot determine with certainty whether the other vessel has the wind on the**

**port or on the starboard side, she shall keep out of the way of the other.**

This rule makes clear that it applies only to two sailing vessels involved in risk of collision. The above actions do not involve power-driven vessels, but when a power-driven vessel sees two sailing vessels then she can determine the course of action taken by either of the sailing vessels.

This part of the rule can easily be understood by the following diagrams (Figure 1).

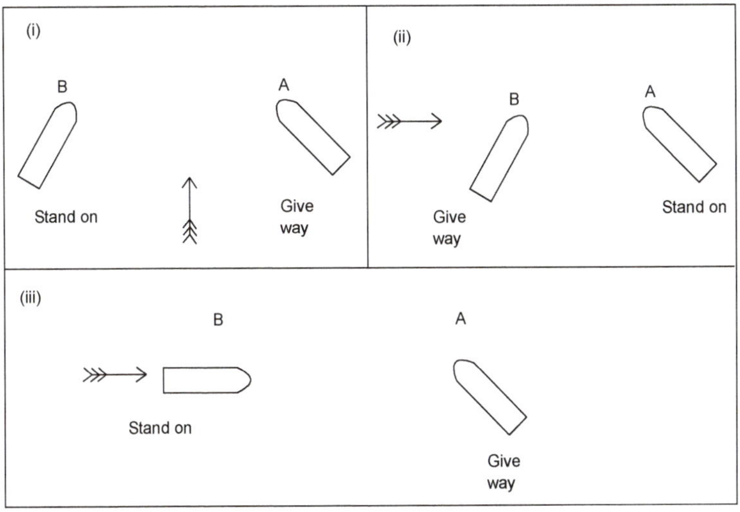


*Figure 1*

**(b)** **For the purposes of this rule the windward side shall be deemed to be the side opposite to that on which the mainsail is carried or, in the case of a square-rigged vessel, the side opposite to that on which the largest fore-and-aft sail is carried.**

# Rule 13

## *Overtaking*

**(a) Notwithstanding anything contained in the rules of Part B, Sections I and II, any vessel overtaking any other shall keep out of the way of the vessel being overtaken.**

This rule overrides all other rules in Sections I and II of Part B. Which means it is the responsibility of the overtaking vessel to keep clear. A point to note here is that this rule overrides all the other rules in Sections I and II, including Rule 18 which involves responsibilities between vessels. When overtaking takes place in clear visibility it is the responsibility of the overtaking vessel to keep clear. While it may be a sailing vessel overtaking a power-driven vessel or a vessel engaged in fishing overtaking a sailing vessel, it remains the responsibility of the overtaking vessel to keep clear.

**(b) A vessel shall be deemed to be overtaking when coming up with another vessel from a direction more than 22.5 degrees abaft her beam, that is, in such a position with reference to the vessel she is overtaking, that at night she would be able to see only the stern light of that vessel but neither of her sidelights.**

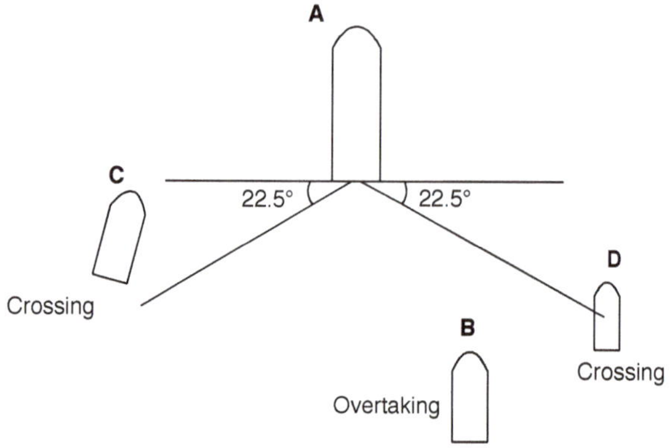


*Figure 2a*

In the above diagram (Figure 2a), B is overtaking A because it is coming up from more than 22.5° abaft the beam. At night, B can initially see A's stern light alone. C is a crossing vessel as she can see a sidelight of A at night. In case of D, the rule states that if a vessel is coming up *more* than 22.5° and vessel D is exactly at 22.5° this makes her a crossing vessel.

**(c) When a vessel is in any doubt as to whether she is overtaking another, she shall assume that this is the case and act accordingly.**

Again, looking at Figure 2a, vessel D explains this part of the rule. At night vessel D might sometimes see vessel A's stern light and sometime vessel A's green sidelight, which will create a doubt. In such a situation vessel D shall assume the responsibility of overtaking and keep clear of vessel A.

**(d) Any subsequent alteration of the bearing between the two vessels shall not make the overtaking vessel a crossing vessel within the meaning of these rules or relieve her of the duty of keeping clear of the overtaken vessel until she is finally past and clear.**

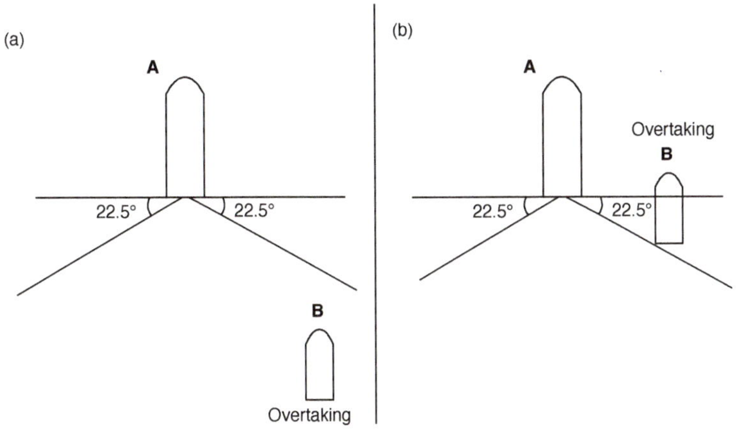


*Figure 2b*

In an overtaking situation when vessel B advances so as to cross the line of two points abaft A's beam, vessel B does not become a crossing vessel. Vessel B remains the overtaking vessel until finally past and clear.

(Please visit www.seamanshiptutor.com/rorbook to download animations on Rule 13.)

## Rule 14

### *Head-on situation*

**(a) When two power-driven vessels are meeting on reciprocal or nearly reciprocal courses so as to involve risk of collision each shall alter her course to starboard so that each shall pass on the port side of the other.**

**(b) Such a situation shall be deemed to exist when a vessel sees the other ahead or nearly ahead and by night she would see the mast head lights of the other in a line or nearly in a line and or both sidelights and by day she observes the corresponding aspect of the other vessel.**

**(c) When a vessel is in any doubt as to whether such a situation exists she shall assume that it does exist and act accordingly.**

Please note this rule concerns two power-driven vessels. Rule 14 differs from other rules in Section II by placing equal responsibility on each power-driven vessel to keep out of each other's way. Both power-driven vessels shall alter course to starboard and pass port of each other. When two power-driven vessels are nearly end-on, a doubt may develop as to whether a head-on situation exists or not. In such cases vessels will assume the case of head-on situation and take action accordingly.

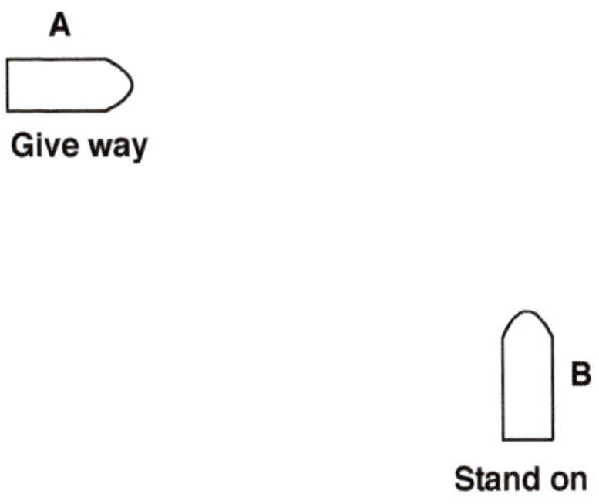


*Figure 3*

## Rule 15

### *Crossing situation*

**When two power-driven vessels are crossing so as to involve risk of collision, the vessel which has the other on her own starboard side shall keep out of the way and shall, if the circumstances of the case admit, avoid crossing ahead of the other vessel.**

Similar to Rule 14, this rule also concerns two power-driven vessels. In the above diagram (Figure 3), power-driven vessel A is the give-way vessel. The latter part of the rule emphasises that the give-way vessel shall avoid crossing ahead of the other vessel. The preferred action under this rule is an alteration of course to starboard to pass astern of the stand-on vessel. One may be to alter course to port and steer parallel to the other vessel's course (which satisfies the avoiding crossing ahead condition) and then reassess the situation.

## Rule 16

### *Action by give-way vessel*

**Every vessel which is directed to keep out of the way of another vessel shall, so far as possible, take early and substantial action to keep well clear.**

Early action means that the other ship will not be panicked into assuming you are not going to alter your course/speed. Substantial means that your manoeuvre is more easily spotted by the ship. As stated in Rule 8, a succession of small alterations is not encouraged as it will not be easily spotted by the other vessel.

## Rule 17

### *Action by stand-on vessel*

**(a) (i) Where one of two vessels is to keep out of the way the other shall keep her course and speed.**

**(ii) The latter vessel may however take action to avoid collision by her manoeuvre alone, as soon as it becomes apparent to her that the vessel required to keep out of the way is not taking appropriate action in compliance with these rules.**

In a close-quarter situation between two vessels, one vessel shall keep out of the way of the other vessel; and the other vessel shall keep her course and speed. This does not mean that the stand-on vessel shall never take action to avoid collision. A stand-on vessel may take action if it is apparent to her that the give-way vessel is not taking appropriate action to avoid collision.

**(b) When, from any cause, the vessel required to keep her course and speed finds herself so close that collision cannot be avoided by the action of the give-way vessel alone, she shall take such action as will best aid to avoid collision.**

When a stand-on vessel finds herself so close to a give-way vessel and determines that collision cannot be avoided by the give-way vessel alone, the stand-on vessel shall take action to avoid the collision.

**(c) A power-driven vessel which takes action in a crossing situation in accordance with sub-paragraph (a)(ii) of this**

**Rule to avoid collision with another power-driven vessel shall, if the circumstances of the case admit, not alter course to port for a vessel on her own port side.**

When a stand-on vessel takes action for a give-way vessel on her port side, the stand-on vessel must not alter course to port side because the give-way vessel may wake up instantly and take action to starboard side in accordance to these rules, which may make the situation even worse. When a vessel, obliged by these rules to stand-on, finds that another vessel is taking insufficient action to avoid collision, then the stand-on vessel should not alter course to port. Implicitly, the stand-on vessel is to alter to starboard, or alter speed. When a stand-on vessel chooses to alter, it should make the alteration large, radical and unmistakable.

(d) **This rule does not relieve the give-way vessel of her obligation to keep out of the way.**

A give-way vessel is not relieved of her obligation of keeping clear under this rule. In other words, a give-way vessel shall not use this rule as an excuse and not take any action in a close-quarter situation.

## Rule 18

### *Responsibilities between vessels*

**Except where Rules 9, 10 and 13 otherwise require:**

(a) **A power-driven vessel underway shall keep out of the way of:**
- (i) **a vessel not under command;**
- (ii) **a vessel restricted in her ability to manoeuvre;**
- (iii) **a vessel engaged in fishing;**
- (iv) **a sailing vessel.**

(b) **A sailing vessel underway shall keep out of the way of:**
- (i) **a vessel not under command;**
- (ii) **a vessel restricted in her ability to manoeuvre;**
- (iii) **a vessel engaged in fishing.**

**(c) A vessel engaged in fishing when underway shall, so far as possible, keep out of the way of:**
**(i) a vessel not under command;**
**(ii) a vessel restricted in her ability to manoeuvre.**

The important point to note here is that this rule does not apply fully in narrow channels (Rule 9) and traffic separation schemes (Rule 10). There are special instructions to sailing vessels, vessels engaged in fishing and vessels of less than 20m in length when operating in TSS and narrow channels; but give-way and stand-on duties will remain as per this rule. In an overtaking situation (Rule 13) this rule is not applicable at all. Otherwise all the above responsibilities are applicable and are self-explanatory.

**(d) (i) Any vessel other than a vessel not under command or a vessel restricted in her ability to manoeuvre shall, if the circumstances of the case admit, avoid impeding the safe passage of a vessel constrained by her draught, exhibiting the signals in Rule 28.**
**(ii) A vessel constrained by her draught shall navigate with particular caution having full regard to her special condition.**

Any vessel other than a vessel not under command or a vessel restricted in her ability to manoeuvre shall avoid impeding the safe passage of the vessel constrained by her draught. This does not make a deep draught vessel a privileged vessel, as she can always reduce her speed if she cannot alter her course because of available depth and width of the water. So she shall navigate with caution having full regard to her special condition.

**(e) A seaplane on the water shall, in general, keep well clear of all vessels and avoid impeding their navigation. In circumstances, however, where risk of collision exists, she shall comply with the rules of this part.**

A seaplane shall avoid impeding the passage of all vessels. If risk of collision exists she shall comply with rules of this part as a power-driven vessel.

*Table B* Impeding

| *Which vessel* | *Shall not/avoid impeding* | *Passage of whom* | *Where* | *Rule* |
|---|---|---|---|---|
| Sailing vessel | Shall not impede | Safe passage of power-driven vessel (PDV)* | TSS | 10(j) |
| Sailing vessel | Shall not impede | Passage of any vessel | Narrow channel | 9(b) |
| Fishing vessel | Shall not impede | Passage of any vessel* | TSS/narrow channel | 9(c)/10(i) |
| Vessel less than 20m | Shall not impede | Safe passage of PDV* | TSS | 10(j) |
| Vessel less than 20m | Shall not impede | Passage of any vessel | Narrow channel | 9(b) |
| Any vessel (other than NUC and RAM) | Avoid impeding | Safe passage of a vessel constrained by her draught (CBD) | Anywhere | 18(d) |
| WIG/seaplane | Avoid impeding | Navigation of any vessel | Anywhere | 18(e, f) |
| A vessel shall not cross narrow channel | If such crossing impedes | The passage of a vessel which can safely navigate only within such a channel or fairway | Within narrow channel | 9(d) |

*Note:* * When following the lane.

(f) (i) **A WIG craft, when taking off, landing and in flight near the surface, shall keep well clear of all other vessels and avoid impeding their navigation;**

(ii) **A WIG craft operating on the water surface shall comply with the rules of this part as a power-driven vessel.**

A WIG craft, in its main operational mode, flies just a few metres above the water surface utilising the surface effect. When taking off, landing and flying near the surface, a WIG craft shall keep clear of all other vessels and avoid impeding their navigation. When operating on the water surface like a normal power-driven vessel, she shall comply with the rules of this part as power-driven vessel.

## SECTION III: CONDUCT OF VESSELS IN RESTRICTED VISIBILITY

### Rule 19

*Conduct of vessels in restricted visibility*

(a) **This rule applies to vessels not in sight of one another when navigating in or near an area of restricted visibility.**

This rule is only applicable in restricted visibility when vessels cannot see each other. As soon as two vessels come in sight of each other this rule does not apply. In restricted visibility, in addition to this rule, Section I of Part B is also applicable.

(b) **Every vessel shall proceed at a safe speed adapted to the prevailing circumstances and conditions of restricted visibility. A power-driven vessel shall have her engines ready for immediate manoeuvre.**

Considering the restricted visibility, every vessel shall proceed at safe speed. A power-driven vessel shall have engines ready for immediate manoeuvre.

(c) Every vessel shall have due regard to the prevailing circumstances and conditions of restricted visibility when complying with the rules of Section I of this part.

(d) A vessel which detects by radar alone the presence of another vessel shall determine if a close-quarters situation is developing and/ or risk of collision exists. If so, she shall take avoiding action in ample time, provided that when such action consists of an alteration of course, so far as possible the following shall be avoided:
- (i) an alteration of course to port for a vessel forward of the beam, other than for a vessel being overtaken;
- (ii) an alteration of course towards a vessel abeam or abaft the beam.

There are a few points to be carefully considered here. This part only applies if a target is detected by radar alone. The part is also only applicable to the targets with risk of collision. The rule obliges one to avoid alteration of course to port for vessel forward of one's own vessel's beam, and to avoid alteration of course towards the vessel abaft of one's own vessel's beam. For ease of understanding, the ways to alter in each case are as follows:

- (i) Alter course to starboard for a vessel forward of the beam.
- (ii) Alter course to starboard for a vessel on or abaft the port beam.
- (iii) Alter course to port for a vessel on or abaft starboard beam.

This can further be simplified by the following situations (Figure 4a–d).

### 4a: Alter course to starboard side (except vessel being overtaken)

(Avoid an alteration of course to port for a vessel forward of the beam, other than for a vessel being overtaken.)

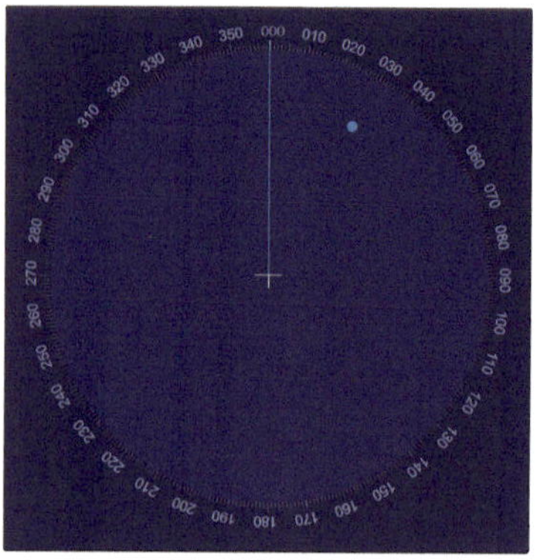


*Figure 4a*

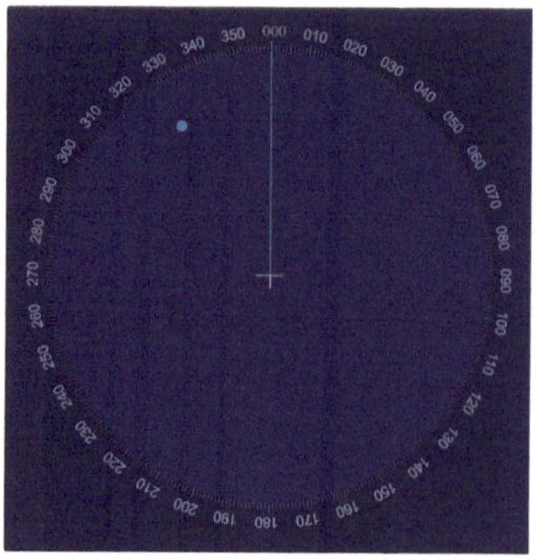


*Figure 4b*

### 4b: Alter course to starboard side (except vessel being overtaken)

(Avoid an alteration of course to port for a vessel forward of the beam, other than for a vessel being overtaken.)

### 4c: Alter course to starboard side

(Avoid an alteration of course towards a vessel abeam or abaft the beam.)

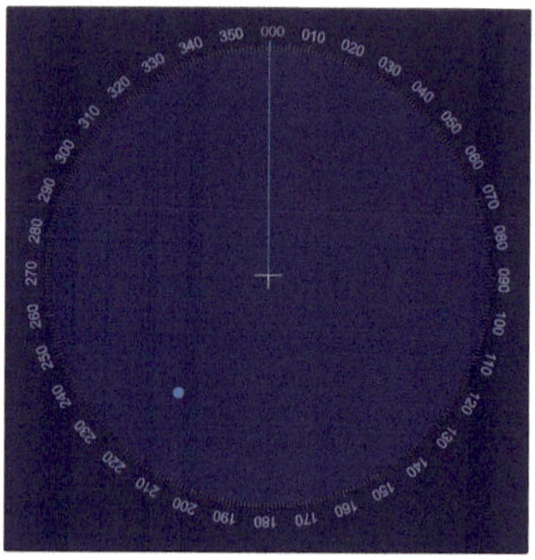


*Figure 4c*

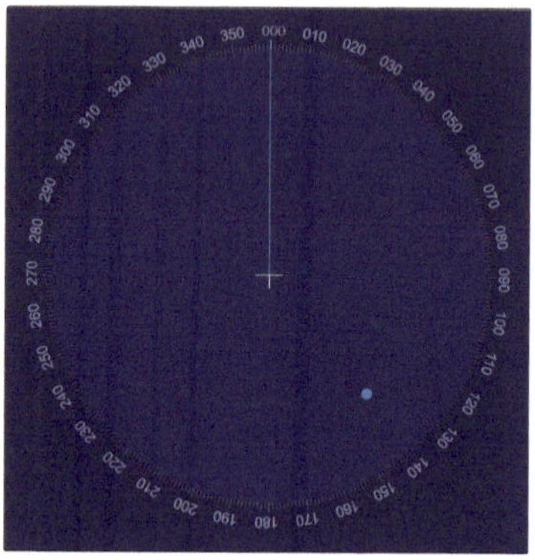


*Figure 4d*

## 4d: Alter course to port side

(Avoid an alteration of course towards a vessel abeam or abaft the beam.)

From the above diagrams it is clear that there is only one case where an alteration of course to port is allowed; that is, when a vessel is on the starboard beam or abaft the starboard beam. Other cases will only accept an alteration of course to starboard side.

All of the above cases can be summed up in the following diagram (Figure 4e).

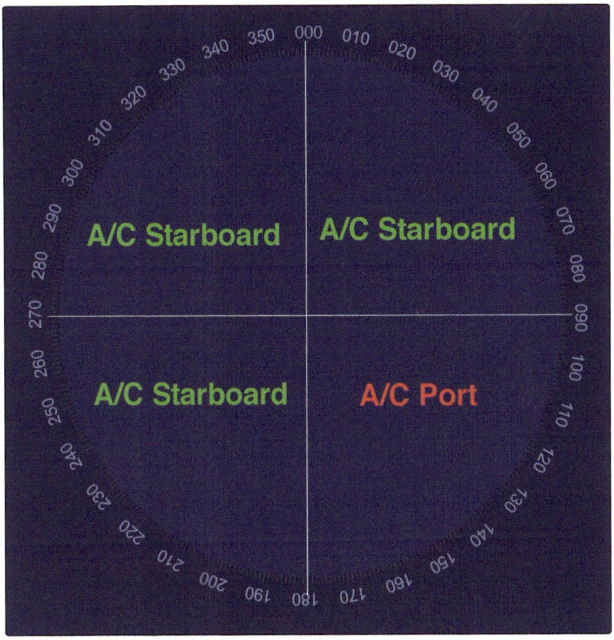


*Figure 4e*

The action shall be taken according to the target position on the radar screen as per the above diagram, other than for a vessel being overtaken. If there is a vessel forward of the beam which is being overtaken, then the overtaking vessel can alter course to either starboard or port in order to overtake the other vessel and avoid collision. Remember, Rule 13 does not apply in restricted visibility.

You must remember that action is only taken for a vessel which is posing a risk of collision. If there are two vessels on the radar screen, one with risk of collision and other with no risk of collision, then action is only to be taken for the vessel with which risk of collision exists. In the following case (Figure 4f), action would be a broad alteration of course to port, since there is no risk of collision with the vessel on port bow.

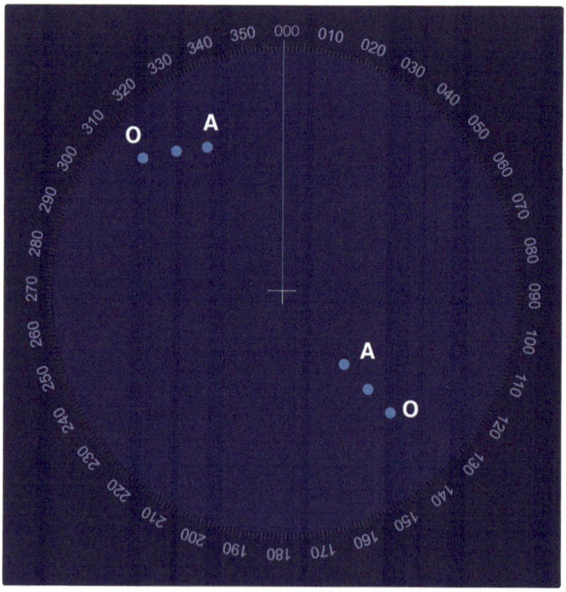


*Figure 4f*

**(e)** **Except where it has been determined that a risk of collision does not exist, every vessel which hears apparently forward of her beam the fog signal of another vessel, or which cannot avoid a close-quarters situation with another vessel forward of her beam, shall reduce her speed to the minimum at which she can be kept on her course. She shall if necessary take all her way off and in any event navigate with extreme caution until danger of collision is over.**

But if the target is not visible on the radar screen and a fog signal is heard forward of beam, that means the target is less than two miles away, which may not be detected by radar, and risk of collision may exist. In this situation a vessel shall reduce her speed to minimum or take all her way off and navigate with extreme caution until the danger of collision has passed.

# Part C
# Lights and shapes

## Rule 20

### *Application*

(a) Rules in this part shall be complied with in all weathers.
(b) The rules concerning lights shall be complied with from sunset to sunrise and during such times no other lights shall be exhibited, except such lights as cannot be mistaken for the lights specified in these rules or do not impair their visibility or distinctive character, or interfere with the keeping of a proper look-out.
(c) The lights prescribed by these rules shall, if carried, also be exhibited from sunrise to sunset in restricted visibility and may be exhibited in all other circumstances when it is deemed necessary.
(d) The rules concerning shapes shall be complied with by day.
(e) The lights and shapes specified in these rules shall comply with the provisions of Annex I to these regulations.

## Rule 21

### *Definitions*

(a) 'Masthead light' means a white light placed over the fore and aft centreline of the vessel showing an unbroken light over an arc of the horizon of 225 degrees and so fixed as to show the light from right ahead to 22.5 degrees abaft the beam on either side of the vessel.

*A Practical Guide to the Rules of the Road, Third edition.* 978-1-138-84389-9.

(b) 'Sidelights' means a green light on the starboard side and a red light on the port side each showing an unbroken light over an arc of the horizon of 112.5 degrees and so fixed as to show the light from the right ahead to 22.5 degrees abaft the beam on its respective side. In a vessel of less than 20 metres in length the sidelights may be combined in one lantern carried on the fore and aft centreline of the vessel.

(c) 'Sternlight' means a white light placed as nearly as practicable at the stern showing an unbroken light over an arc of the horizon of 135 degrees and so fixed as to show the light 67.5 degrees from right aft on each side of the vessel.

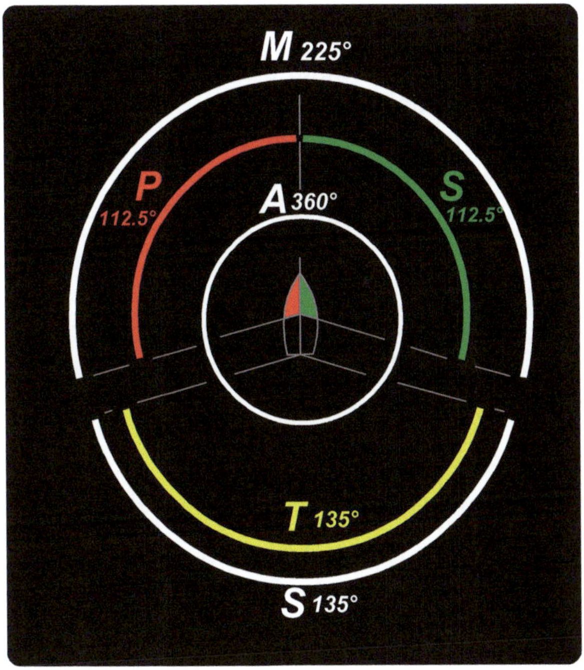


*Figure 5*

(d) 'Towing light' means a yellow light having the same characteristics as the 'sternlight' defined in paragraph (c) of this rule.

(e) 'All-round light' means a light showing an unbroken light over an arc of the horizon of 360 degrees.

(f) 'Flashing light' means a light flashing at regular intervals at a frequency of 120 flashes or more per minute.

## Rule 22

### *Visibility of lights*

The lights prescribed in these rules shall have an intensity as specified in Section 8 of Annex I to these regulations so as to be visible at the following minimum ranges:

(a) In vessels of 50 metres or more in length:

- a masthead light, 6 miles;
- a sidelight, 3 miles;
- a stern light, 3 miles;
- a towing light, 3 miles;
- a white, red, green or yellow all-round light, 3 miles.

(b) In vessels of 12 metres or more in length but less than 50 metres in length:

- a masthead light, 5 miles; except that where the length of the vessel is less than 20 metres, 3 miles;
- a sidelight, 2 miles;
- a sternlight, 2 miles;
- a towing light, 2 miles;
- a white, red, green or yellow all-round light, 2 miles.

(c) In vessels of less than 12 metres in length:

- a masthead light, 2 miles;
- a sidelight, 1 mile;
- a sternlight, 2 miles;
- a towing light, 2 miles
- a white, red, green or yellow all-round light, 2 miles.

(d) In inconspicuous, partly submerged vessels or objects being towed:
- a white all-round light, 3 miles.

## Rule 23

### *Power-driven vessels underway*

(a) A power-driven vessel underway shall exhibit:
  (i) a masthead light forward;
  (ii) a second masthead light abaft of and higher than the forward one; except that a vessel of less than 50 metres in length shall not be obliged to exhibit such light but may do so;
  (iii) sidelights;
  (iv) a sternlight.

(b) An air-cushion vessel when operating in the non-displacement mode shall, in addition to the lights prescribed in paragraph (a) of this rule, exhibit an all-round flashing yellow light.

(c) A WIG craft only when taking off, landing and in flight near the surface shall, in addition to the lights prescribed in paragraph (a) of this rule, exhibit a high-intensity all-round flashing red light.

(d) (i) A power-driven vessel of less than 12 metres in length may in lieu of the lights prescribed in paragraph (a) of this rule exhibit an all-round white light and sidelights;
  (ii) a power-driven vessel of less than 7 metres in length whose maximum speed does not exceed 7 knots may in lieu of the lights prescribed in paragraph (a) of this rule exhibit an all-round white light and shall, if practicable, also exhibit sidelights;
  (iii) the masthead light or all-round white light on a power-driven vessel of less than 12 metres in length may be displaced from the fore and aft centre line of the vessel if centreline fitting is not practicable,

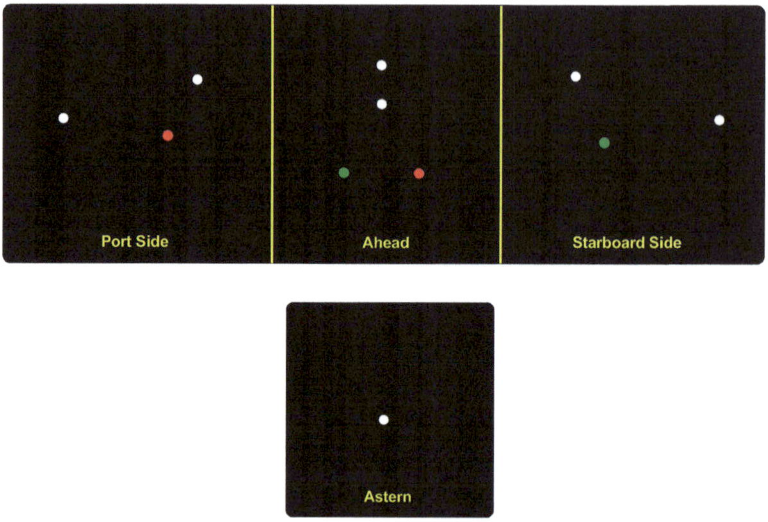


*Figure 6a* Power-driven vessel, probably 50m or more in length, underway

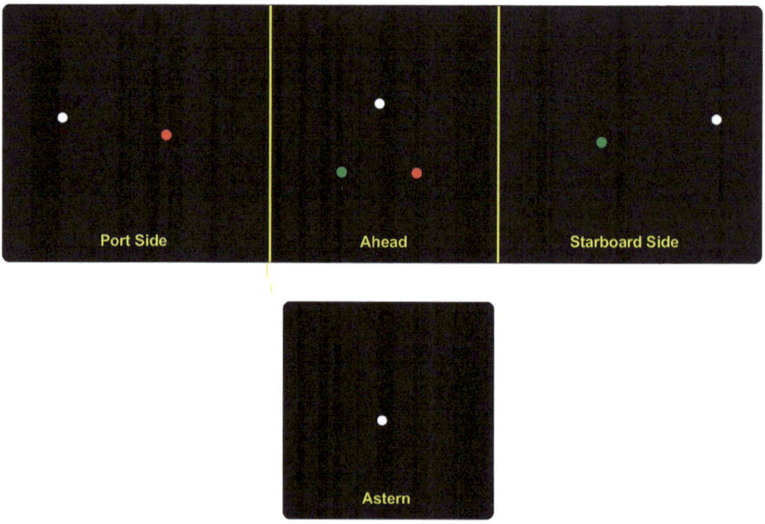


*Figure 6b* Power-driven vessel, less than 50m in length, underway

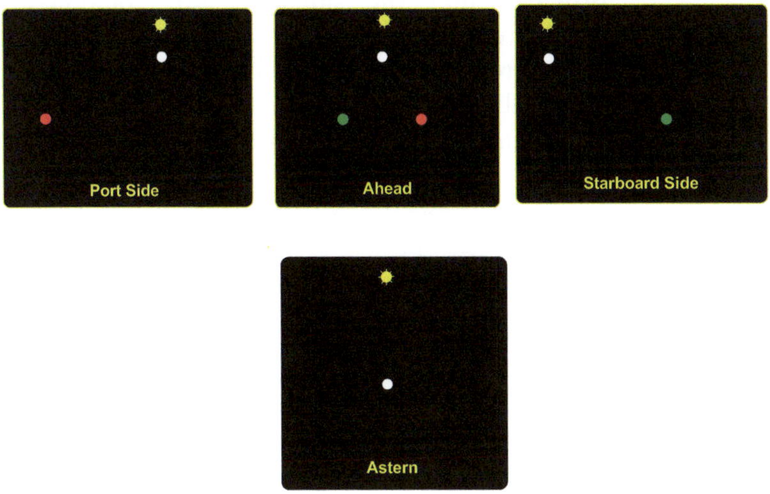


*Figure 6c* Air-cushion vessel, underway

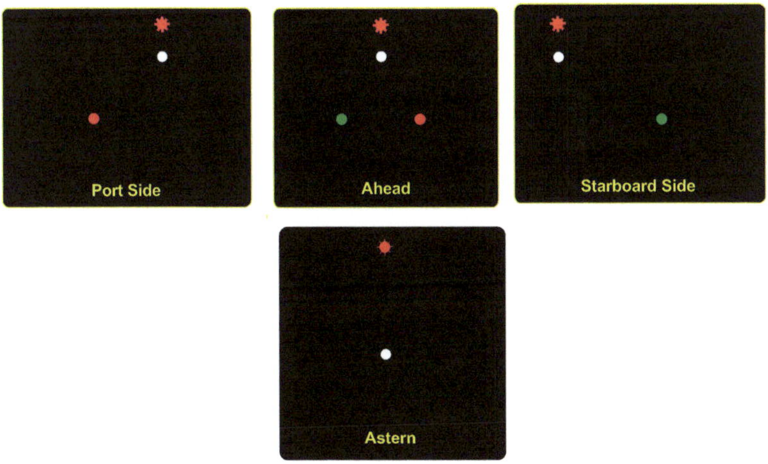


*Figure 6d* WIG craft, underway (in WIG mode)

provided that the sidelights are combined in one lantern which shall be carried on the fore and aft centre line of the vessel or located as nearly as practicable in the same fore and aft line as the masthead light or the all-round white light.

## Rule 24

### *Towing and pushing*

(a) A power-driven vessel when towing shall exhibit:
  (i) instead of the light prescribed in Rule 23(a)(i) or (a)(ii), two masthead lights in a vertical line. When the length of the tow, measuring from the stern of the towing vessel to the after end of the tow exceeds 200 metres, three such lights in a vertical line;
  (ii) sidelights;
  (iii) a sternlight;
  (iv) a towing light in a vertical line above the sternlight;
  (v) when the length of the tow exceeds 200 metres, a diamond shape where it can best be seen.

(b) When a pushing vessel and a vessel being pushed ahead are rigidly connected in a composite unit they shall be regarded as a power-driven vessel and exhibit the lights prescribed in Rule 23.

(c) A power-driven vessel when pushing ahead or towing alongside, except in the case of a composite unit, shall exhibit:
  (i) instead of the light prescribed in Rule 23(a)(i) or (a)(ii), two masthead lights in a vertical line;
  (ii) sidelights;
  (iii) a sternlight.

(d) A power-driven vessel to which paragraph (a) or (c) of this rule applies shall also comply with Rule 23(a) (ii).

(e) A vessel or object being towed, other than those mentioned in paragraph (g) of this Rule, shall exhibit:
  (i) sidelights;
  (ii) a sternlight;

(iii) when the length of the tow exceeds 200 metres, a diamond shape where it can best be seen.

(f) Provided that any number of vessels being towed alongside or pushed in a group shall be lighted as one vessel,
   (i) a vessel being pushed ahead, not being part of a composite unit, shall exhibit at the forward end sidelights;
   (ii) a vessel being towed alongside shall exhibit a sternlight and at the forward end, sidelights.

(g) An inconspicuous, partly submerged vessel or object, or combination of such vessels or objects being towed, shall exhibit:
   (i) if it is less than 25 metres in breadth, one all-round white light at or near the forward end and one at or near the after end except that dracones need not exhibit a light at or near the forward end;
   (ii) if it is 25 metres or more in breadth, two additional all-round white lights at or near the extremities of its breadth;
   (iii) if it exceeds 100 metres in length, additional all-round white lights between the lights prescribed in subparagraphs (i) and (ii) so that the distance between the lights shall not exceed 100 metres;
   (iv) a diamond shape at or near the aftermost extremity of the last vessel or object being towed and if the length of the tow exceeds 200 metres an additional diamond shape where it can best be seen and located as far forward as is practicable.

(h) Where from any sufficient cause it is impracticable for a vessel or object being towed to exhibit the lights or shapes prescribed in paragraph (e) or (g) of this rule, all possible measures shall be taken to light the vessel or object towed or at least to indicate the presence of such vessel or object.

(i) Where from any sufficient cause it is impracticable for a vessel not normally engaged in towing operations to display the lights prescribed in paragraph (a) or (c) of this rule, such vessel shall not be required to exhibit those lights when engaged in towing another vessel in distress or otherwise in need of assistance. All possible measures shall be taken to indicate the nature of the relationship between the towing vessel and the vessel being towed as authorized by Rule 36, in particular by illuminating the towline.

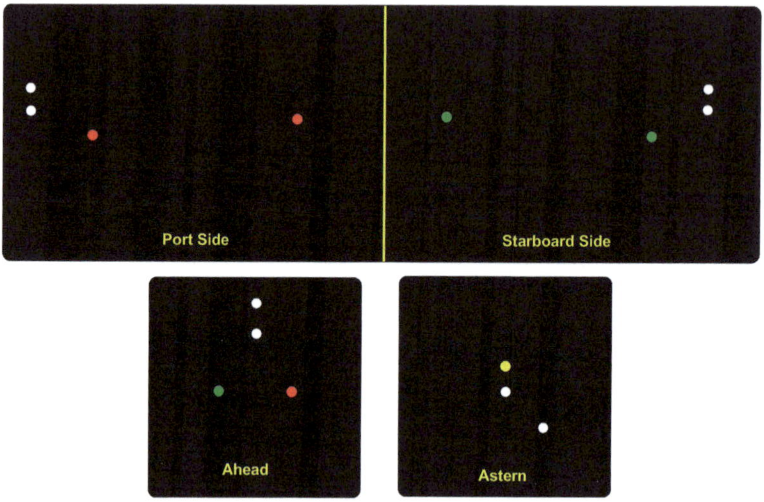


*Figure 7a* Power-driven vessel less than 50m in length, towing, length of tow not more than 200m, underway

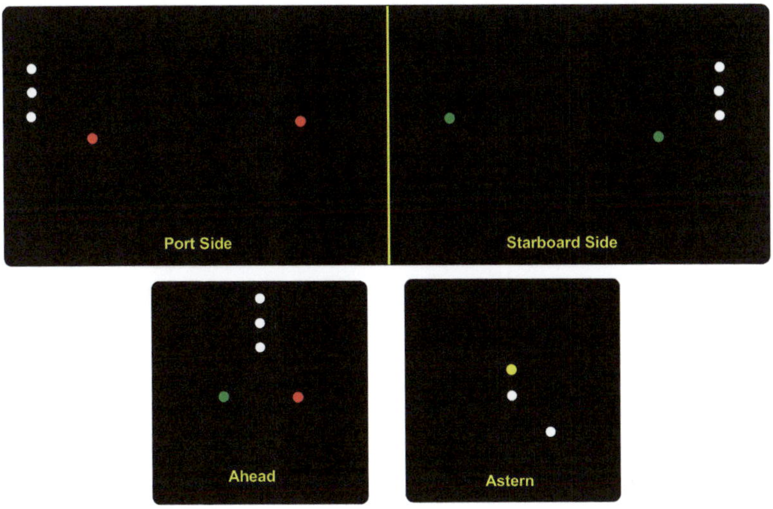


*Figure 7b* Power-driven vessel less than 50m in length, towing, length of tow more than 200m, underway

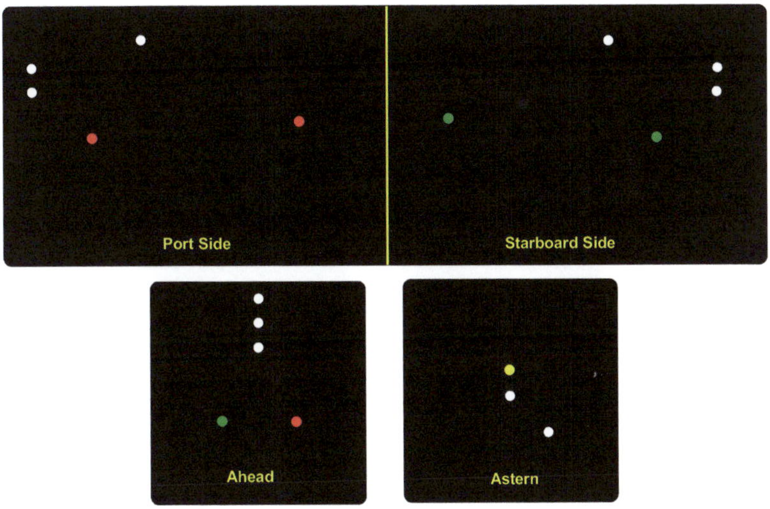


*Figure 7c* Power-driven vessel probably 50m or more in length, towing, length of tow less than 200m, underway

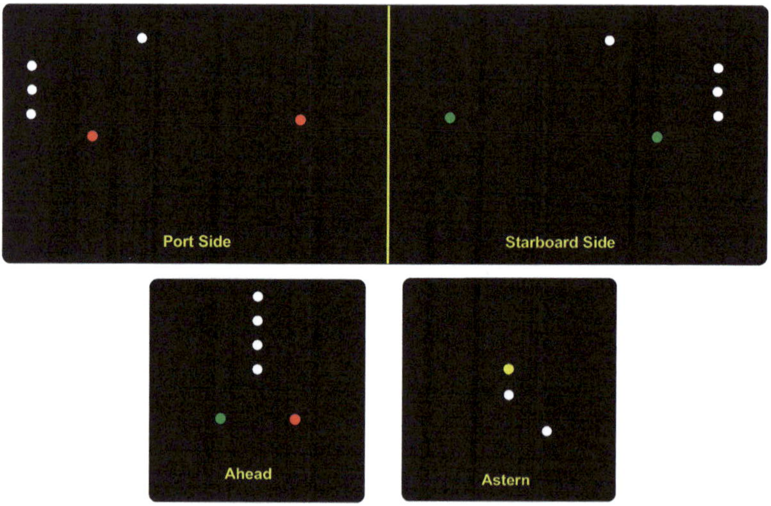


*Figure 7d* Power-driven vessel probably 50m or more in length, towing, length of tow more than 200m, underway

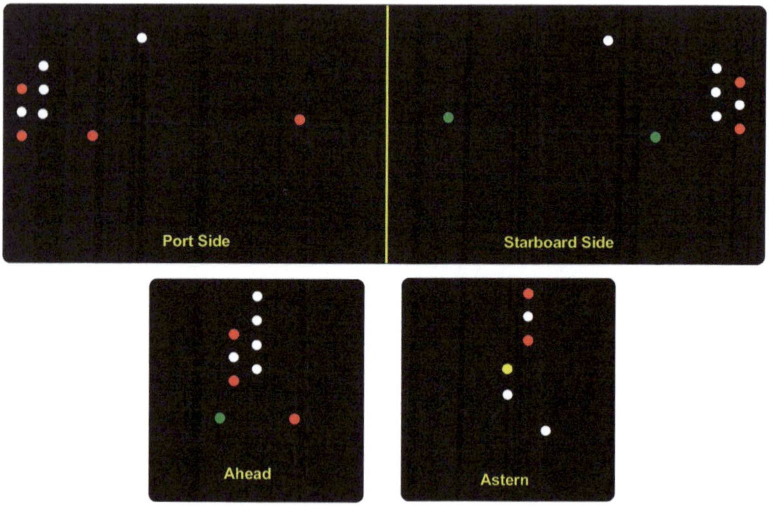


*Figure 7e* Power-driven vessel probably 50m or more in length, towing, length of tow more than 200m, restricted in their ability to manoeuvre, underway

*Figure 7f* Power-driven vessel less than 50m in length, towing, length of tow more than 200m, underway. Partly submerged object: length less than 100m, breadth less than 25m

*Figure 7g* Power-driven vessel less than 50m in length, towing, length of tow more than 200m, underway. Partly submerged object: length more than 100m (less than 200m), breadth less than 25m

*Figure 7h* Power-driven vessel less than 50m in length, towing, length of tow more than 200m, underway. Partly submerged object: length less than 100m, breadth more than 25m

*Figure 7i* Power-driven vessel less than 50m in length, towing, length of tow more than 200m, underway. Partly submerged object: length more than 100m (less than 200m), breadth more than 25m

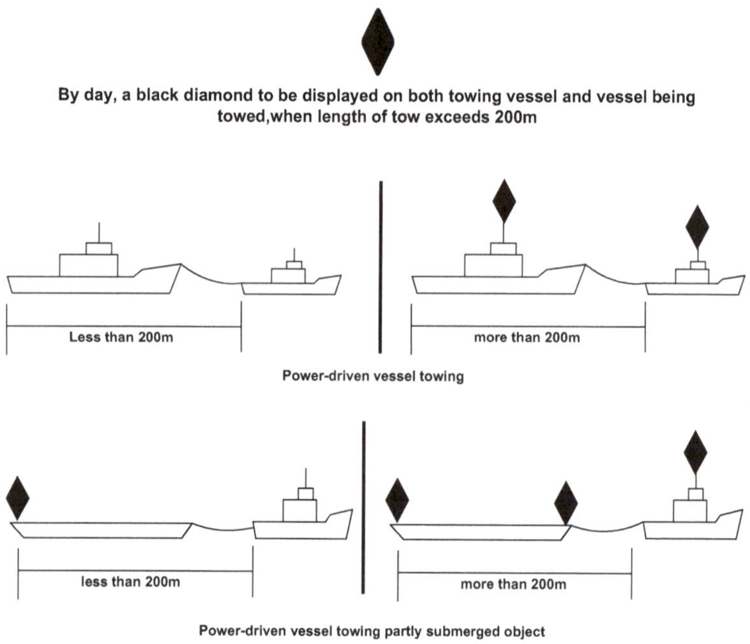


*Figure 7j*

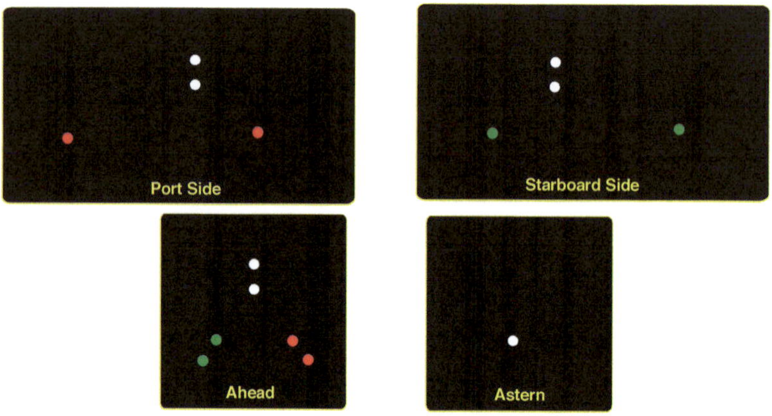


*Figure 7k* Power-driven vessel less than 50m in length, pushing ahead, underway

## Rule 25

### *Sailing vessels underway and vessels under oars*

(a) A sailing vessel underway shall exhibit:
  (i) sidelights;
  (ii) a sternlight.

(b) In a sailing vessel of less than 20 metres in length the lights prescribed in paragraph (a) of this rule may be combined in one lantern carried at or near the top of the mast where it can best be seen.

(c) A sailing vessel underway may, in addition to the lights prescribed in paragraph (a) of this rule, exhibit at or near the top of the mast, where they can best be seen, two all-round lights in a vertical line, the upper being red and the lower green, but these lights shall not be exhibited in conjunction with the combined lantern permitted by paragraph (b) of this rule.

(d) (i) A sailing vessel of less than 7 metres in length shall, if practicable, exhibit the lights prescribed in paragraph (a) or (b) of this rule, but if she does not, she shall have ready at hand an electric torch or lighted lantern showing a white light which shall be exhibited in sufficient time to prevent collision.
  (ii) A vessel under oars may exhibit the lights prescribed in this rule for sailing vessels, but if she does not, she shall have ready at hand an electric torch or lighted lantern showing a white light which shall be exhibited in sufficient time to prevent collision.

(e) A vessel proceeding under sail when also being propelled by machinery shall exhibit forward where it can best be seen a conical shape, apex downwards.

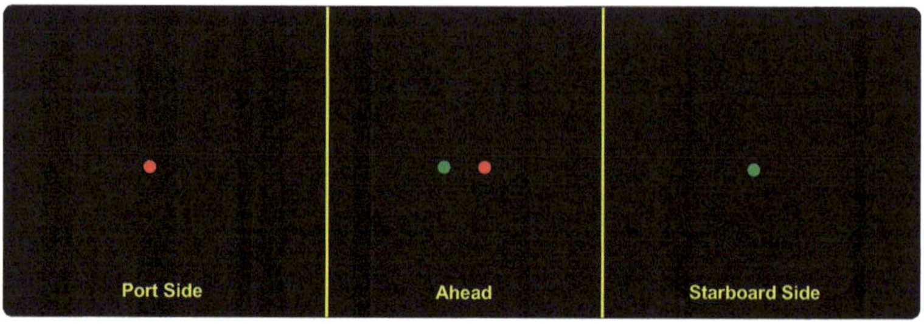


*Figure 8a* Sailing vessel, not showing optional sailing lights, underway

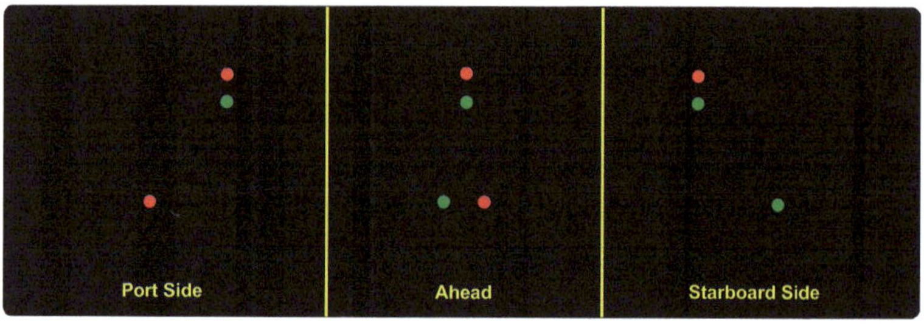


*Figure 8b* Sailing vessel, showing optional all-round sailing lights, underway

## Rule 26

### *Fishing vessels*

(a) A vessel engaged in fishing, whether underway or at anchor, shall exhibit only the lights and shapes prescribed in this rule.

(b) A vessel when engaged in trawling, by which is meant the dragging through the water of a dredge net or other apparatus used as a fishing appliance, shall exhibit:

  (i) two all-round lights in a vertical line, the upper being green and the lower white, or a shape consisting of two cones with their apexes together in a vertical line one above the other;

  (ii) a masthead light abaft of and higher than the all-round green light; a vessel of less than 50 metres in length shall not be obliged to exhibit such a light but may do so;

(iii) when making way through the water, in addition to the lights prescribed in this paragraph, sidelights and a stern light.

(c) A vessel engaged in fishing, other than trawling, shall exhibit:

(i) two all-round lights in a vertical line, the upper being red and the lower white, or a shape consisting of two cones with apexes together in a vertical line one above the other;

(ii) when there is outlying gear extending more than 150 metres horizontally from the vessel, an all round white light or a cone apex upwards in the direction of the gear;

(iii) when making way through the water, in addition to the lights prescribed in this paragraph, sidelights and a sternlight.

(d) The additional signals described in Annex II to these regulations apply to a vessel engaged in fishing in close proximity to other vessels engaged in fishing.

(e) A vessel when not engaged in fishing shall not exhibit the lights or shapes prescribed in this rule, but only those prescribed for a vessel of her length.

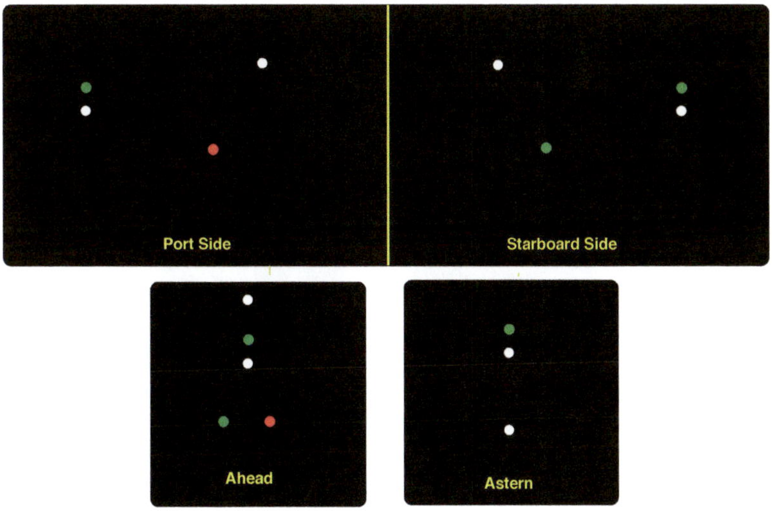


*Figure 9a* Vessel engaged in trawling, probably 50m or more in length, making way through the water

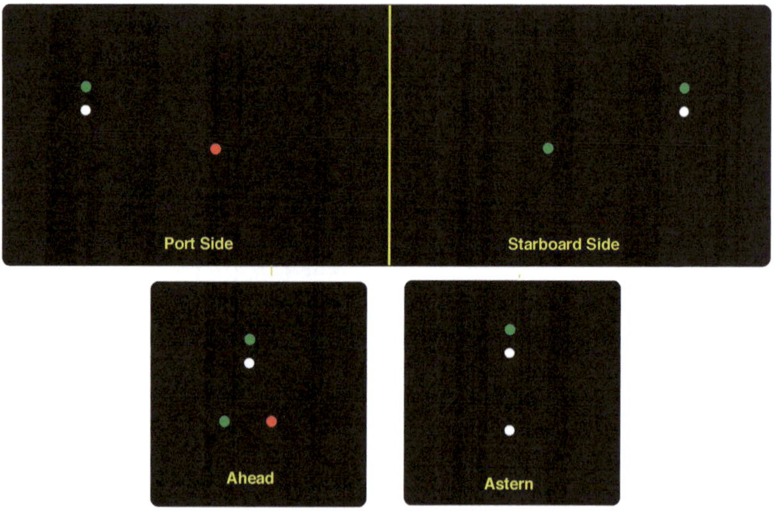


*Figure 9b* Vessel engaged in trawling, less than 50m in length, making way through the water

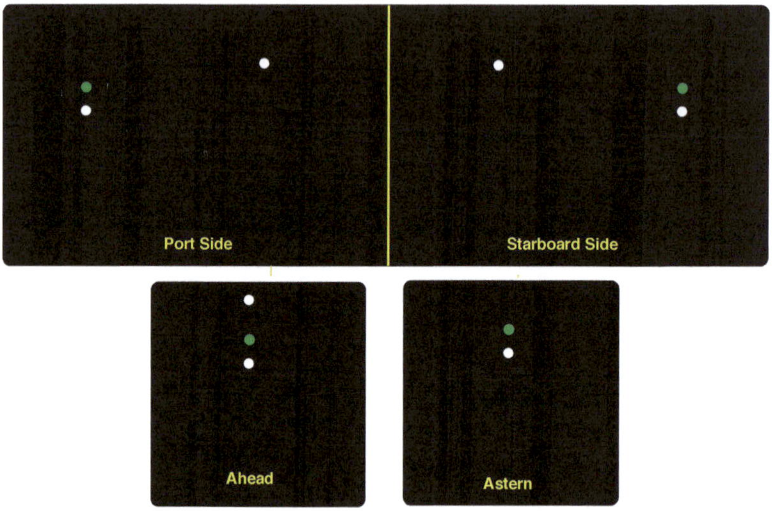


*Figure 9c* Vessel engaged in trawling, probably 50m or more in length, not making way through the water

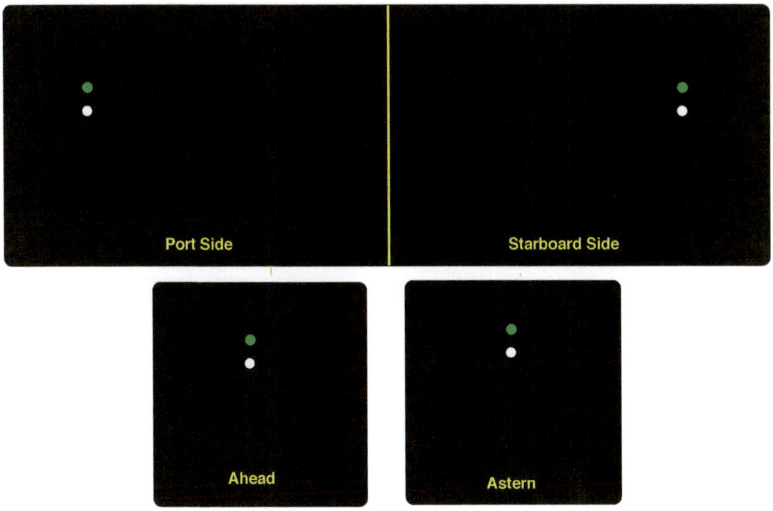


*Figure 9d* Vessel engaged in trawling, less than 50m in length, not making way through the water

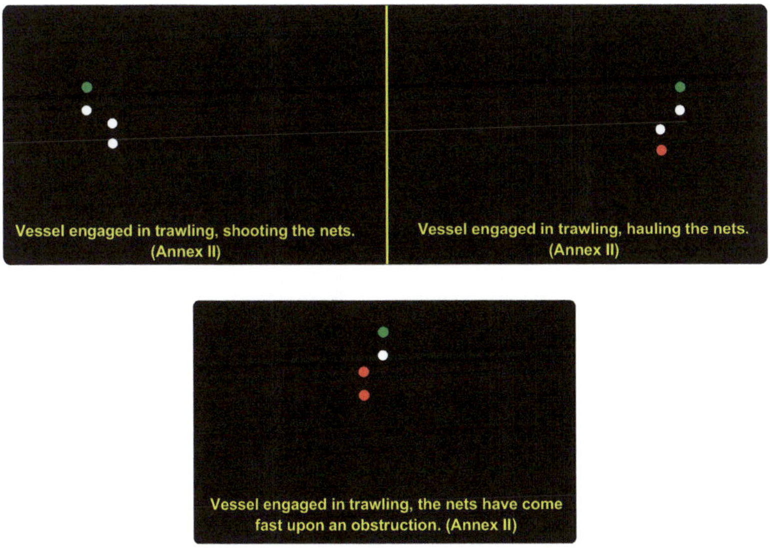


*Figure 9e* Vessel engaged in trawling, less than 50m in length, not making way through the water

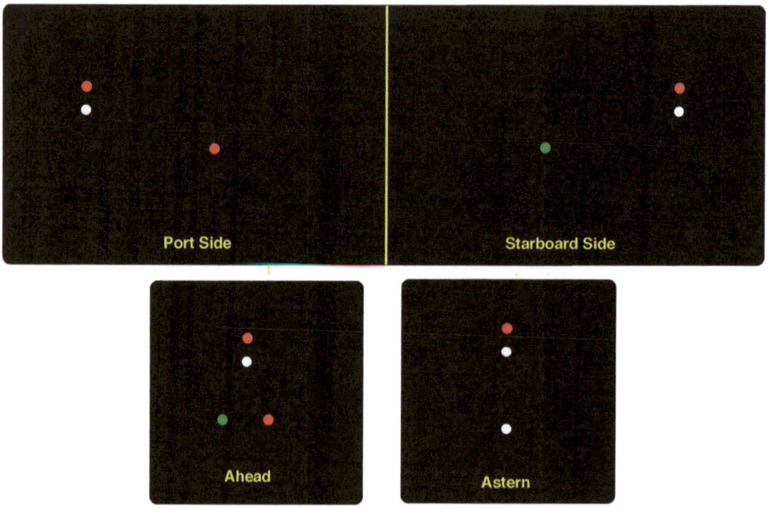


*Figure 9f* Vessel engaged in fishing, other than trawling, making way through the water

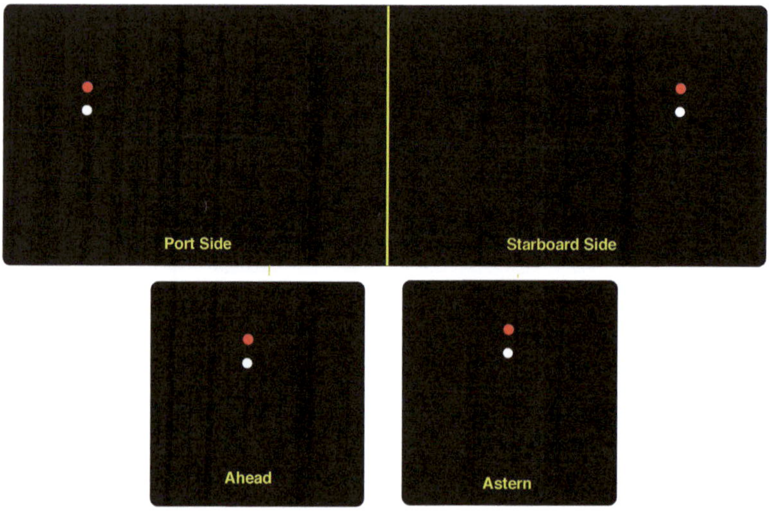


*Figure 9g* Vessel engaged in fishing, other than trawling, not making way through the water

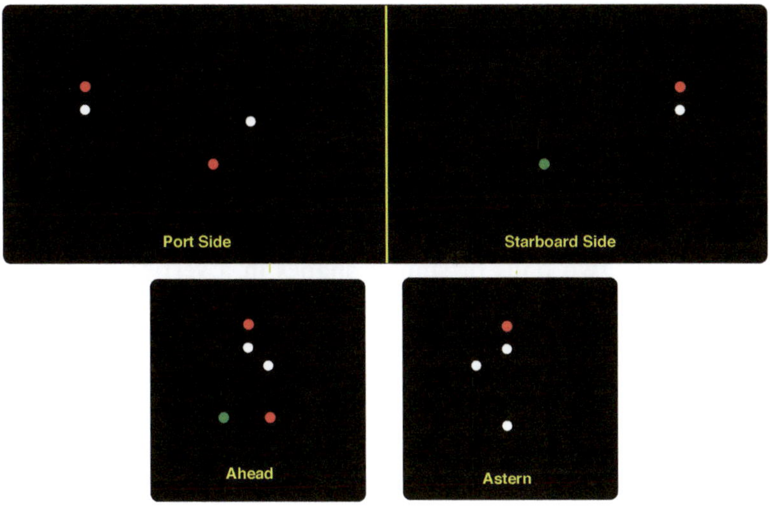


*Figure 9h* Vessel engaged in fishing, other than trawling, outlying gear extending more than 150m from the vessel on her port side, making way through the water

*Figure 9i* Two black cones apex together to be displayed by day

A cone apex upwards in the direction of the gears extending more than 150m in length

## Rule 27

*Vessels not under command or restricted in their ability to manoeuvre*

(a) A vessel not under command shall exhibit:
- (i) two all-round red lights in a vertical line where they can best be seen;
- (ii) two balls or similar shapes in a vertical line where they can best be seen;
- (iii) when making way through the water, in addition to the lights prescribed in this paragraph, sidelights and a stern light.

(b) A vessel restricted in her ability to manoeuvre, except a vessel engaged in mine-clearance operations, shall exhibit:
- (i) three all-round lights in a vertical line where they can best be seen. The highest and lowest of these lights shall be red and the middle light shall be white;
- (ii) three shapes in a vertical line where they can best be seen. The highest and lowest of these shapes shall be balls and the middle one a diamond;
- (iii) when making way through the water, a masthead light or lights, sidelights and a sternlight, in addition to the lights prescribed in sub-paragraph (i);
- (iv) when at anchor, in addition to the lights or shapes prescribed in sub-paragraphs (i) and (ii), the light, lights or shape prescribed in Rule 30.

(c) A power-driven vessel engaged in a towing operation such as severely restricts the towing vessel and her tow in their ability to deviate from their course shall, in addition to the lights or shapes prescribed in Rule 24(a), exhibit the lights or shapes prescribed in sub-paragraphs (b)(i) and (ii) of this rule.

(d) A vessel engaged in dredging or underwater operations, when restricted in her ability to manoeuvre, shall exhibit the lights and shapes prescribed in sub-paragraphs (b)(i), (ii) and (iii) of this rule and shall in addition, when an obstruction exists, exhibit:

(i) two all-round red lights or two balls in a vertical line to indicate the side on which the obstruction exists;
(ii) two all-round green lights or two diamonds in a vertical line to indicate the side on which another vessel may pass;
(iii) when at anchor, the lights or shapes prescribed in this paragraph instead of the lights or shape prescribed in Rule 30.

(e) Whenever the size of a vessel engaged in diving operations makes it impracticable to exhibit all lights and shapes prescribed in paragraph (d) of this rule, the following shall be exhibited:
(i) three all-round lights in a vertical line where they can best be seen. The highest and lowest of these lights shall be red and the middle light shall be white;
(ii) a rigid replica of the International Code flag 'A' not less than 1 metre in height. Measures shall be taken to ensure its all-round visibility.

(f) A vessel engaged in mine-clearance operations shall in addition to the lights prescribed for a power-driven vessel in Rule 23 or to the lights or shape prescribed for a vessel at anchor in Rule 30 as appropriate, exhibit three all-round green lights or three balls. One of these lights or shapes shall be exhibited near the foremast head and one at each end of the fore yard. These lights or shapes indicate that it is dangerous for another vessel to approach within 1,000 metres of the mine-clearance vessel.

(g) Vessels of less than 12 metres in length, except those engaged in diving operations, shall not be required to exhibit the lights and shapes prescribed in this rule.

(h) The signals prescribed in this rule are not signals of vessels in distress and requiring assistance. Such signals are contained in Annex IV to these regulations.

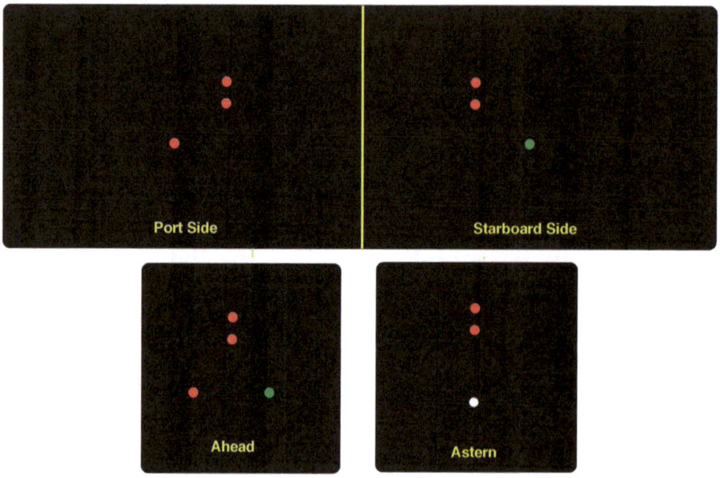


*Figure 10a* Vessel not under command, making way through the water

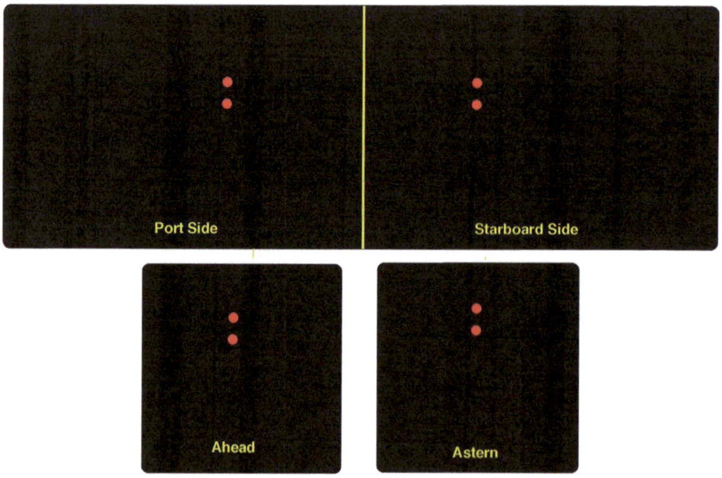


*Figure 10b* Vessel not under command, not making way through the water

*Figure 10c* Two black balls in a vertical line to be displayed by day by a vessel not under command

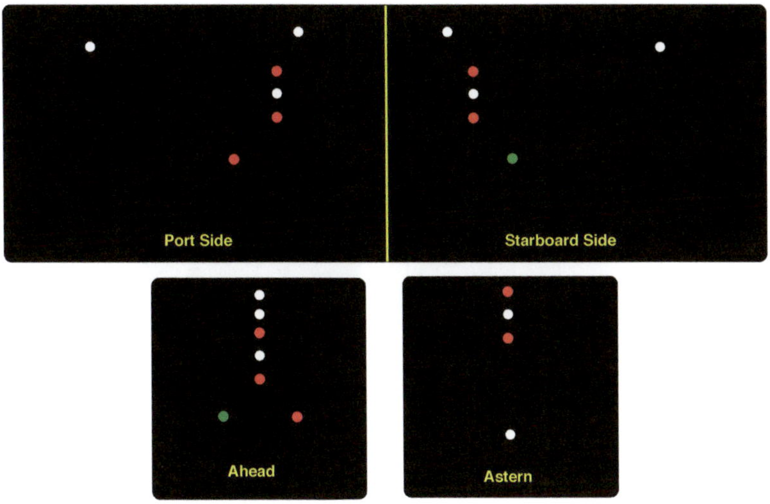


*Figure 10d* Vessel restricted in her ability to manoeuvre, probably 50m or more in length, making way through the water

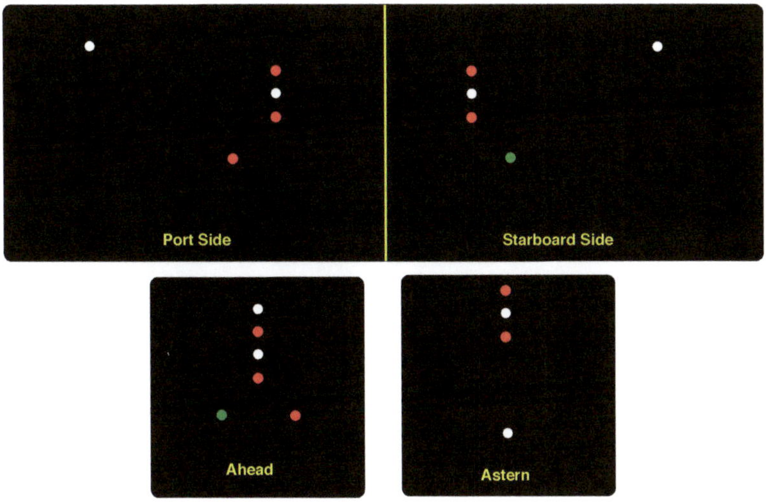


*Figure 10e* Vessel restricted in her ability to manoeuvre, less than 50m in length, making way through the water

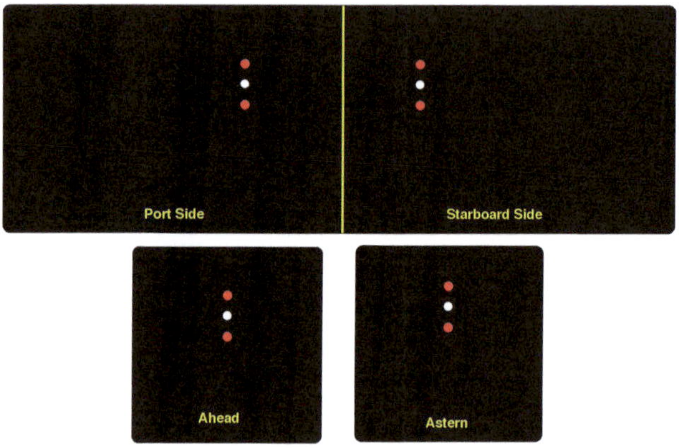


*Figure 10f* Vessel restricted in her ability to manoeuvre, not making way through the water

*Figure 10g* A vessel restricted in her ability to manoeuvre shall, by day, exhibit three shapes in the vertical line, upper and lower being black balls and a black diamond in the middle

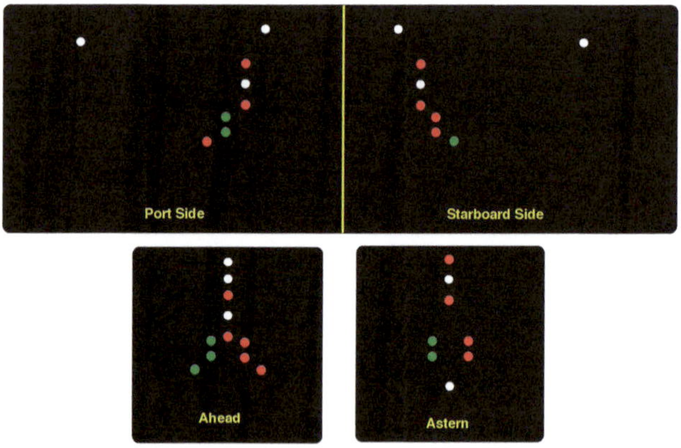


*Figure 10h* Vessel engaged in dredging operations, probably 50m or more in length, restricted in her ability to manoeuvre, making way through the water

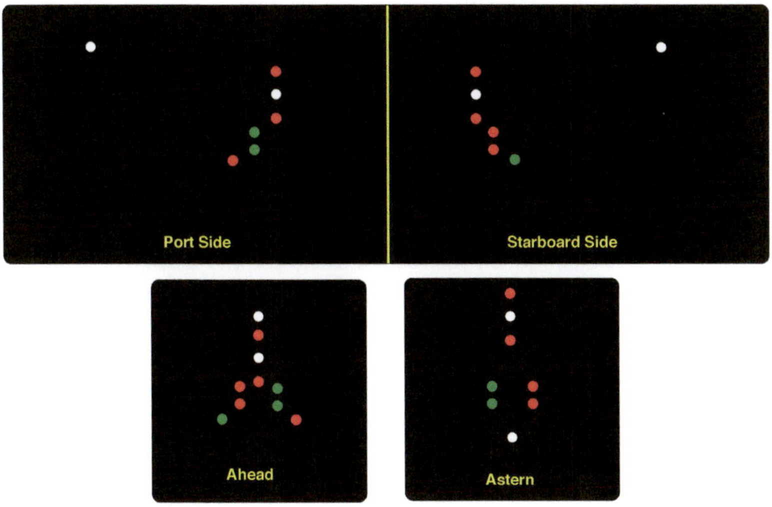


*Figure 10i* Vessel engaged in dredging operations, less than 50m in length, restricted in her ability to manoeuvre, making way through the water

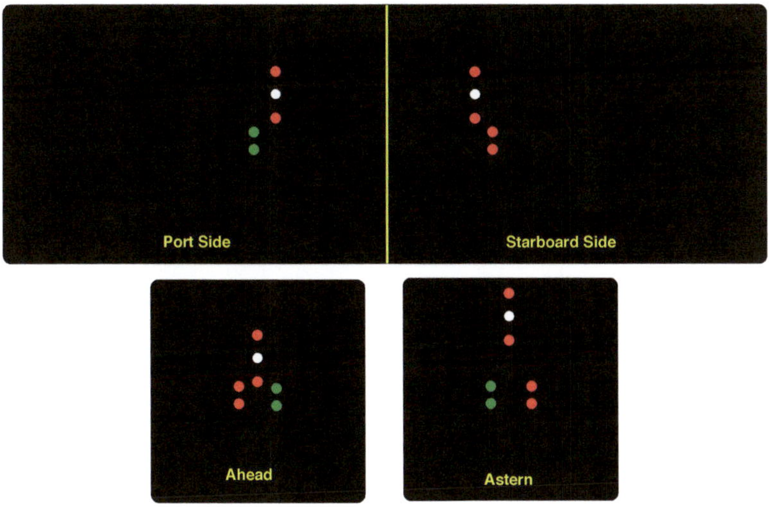


*Figure 10j* Vessel engaged in dredging operations, restricted in her ability to manoeuvre, not making way through the water

*Figure 10k* A vessel engaged in dredging or underwater operations, when restricted in her ability to manoeuvre, in addition to being restricted in her ability to manoeuvre shapes, shall exhibit two black balls in a vertical line to indicate the side on which obstruction exists and two black diamonds in a vertical line to indicate the side which is clear and another vessel may pass

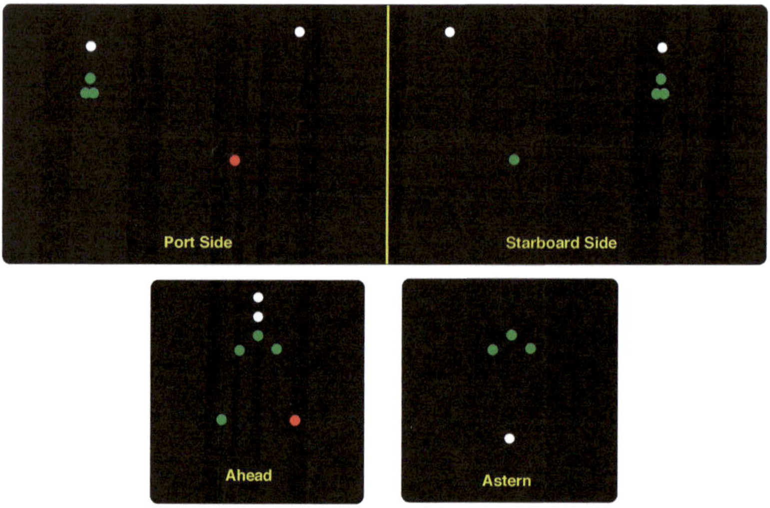


*Figure 10l* Vessel engaged in mine-clearance operations, probably 50m or more in length, underway

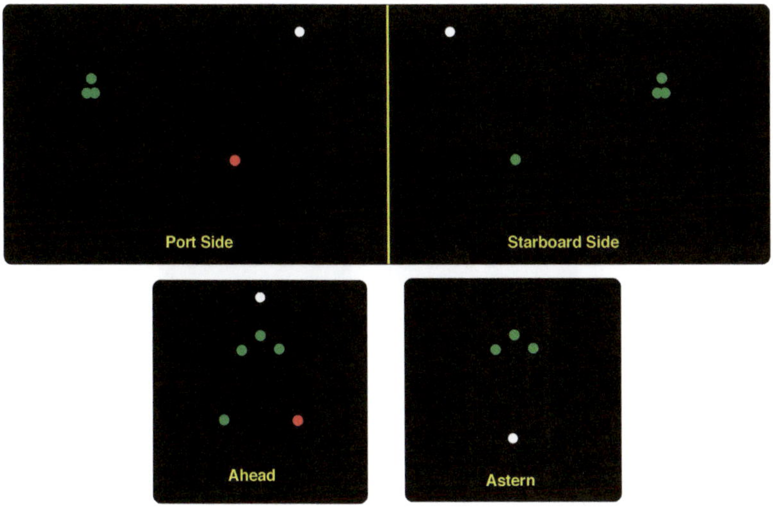


*Figure 10m* Vessel engaged in mine-clearance operations, less than 50m in length, underway

*Figure 10n* By day, a vessel engaged in mine-clearance operations, shall exhibit three black balls. One of these balls shall be exhibited near the foremast head and one at each end of the fore yard

## Rule 28

### *Vessels constrained by their draught*

A vessel constrained by her draught may, in addition to the lights prescribed for power-driven vessels in Rule 23, exhibit where they can best be seen three all-round red lights in a vertical line, or a cylinder.

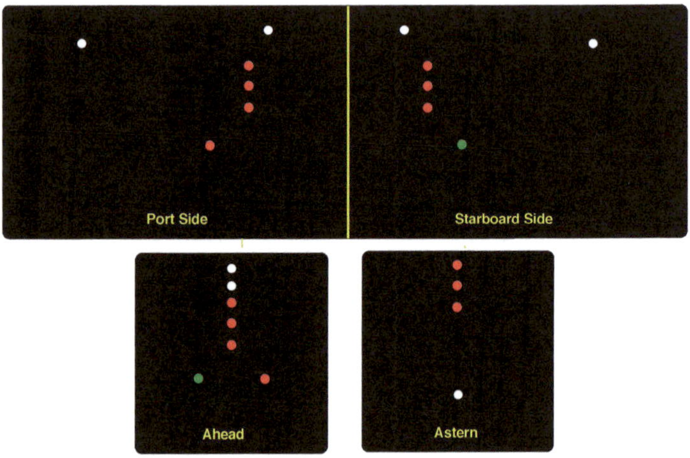


*Figure 11a* Vessel constrained by her draught, probably 50m or more in length, underway

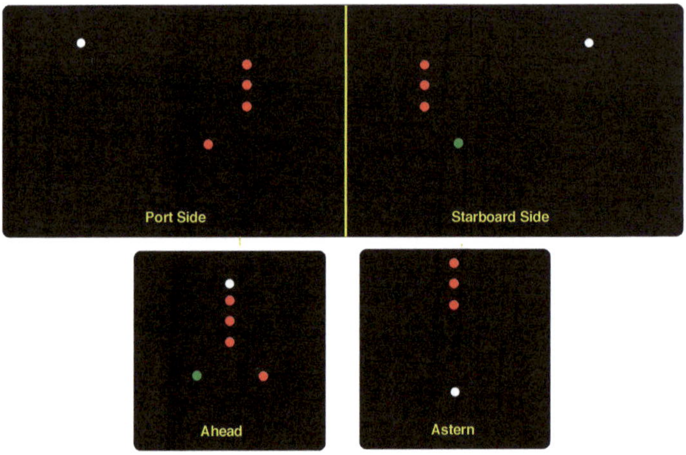


*Figure 11b* Vessel constrained by her draught, less than 50m in length, underway

*Figure 11c* A vessel constrained by her draught shall, by day, exhibit a black cylinder

# Rule 29

## *Pilot vessels*

(a) A vessel engaged on pilotage duty shall exhibit:

(i) at or near the masthead, two all-round lights in a vertical line, the upper being white and the lower red;

(ii) when underway, in addition, sidelights and a sternlight;

(iii) when at anchor, in addition to the lights prescribed in sub-paragraph (i), the light, lights or shape prescribed in Rule 30 for vessels at anchor.

(b) A pilot vessel when not engaged on pilotage duty shall exhibit the lights or shapes prescribed for a similar vessel of her length.

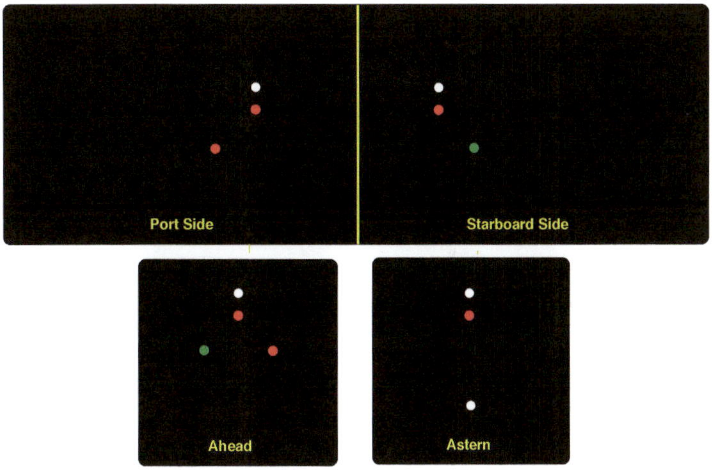


*Figure 12a* Vessel engaged in pilotage duties, underway

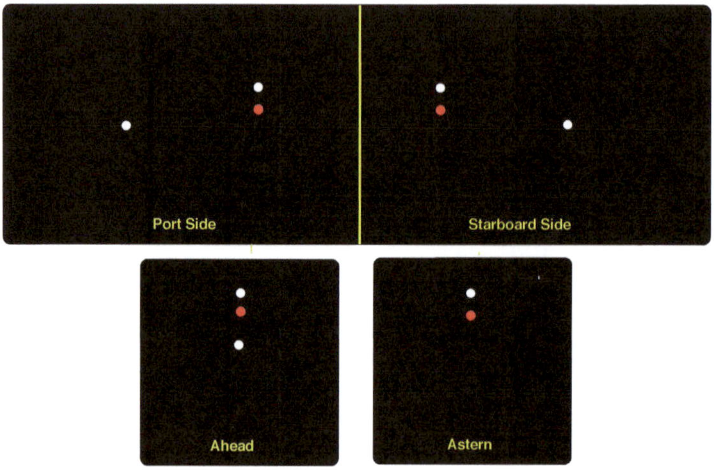


*Figure 12b* Vessel engaged in pilotage duties, less than 50m in length, at anchor

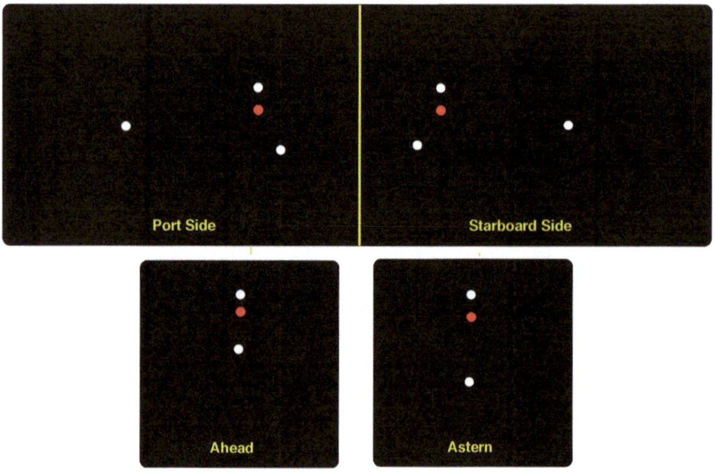


*Figure 12c* Vessel engaged in pilotage duties, probably 50m or more in length, at anchor

## Rule 30

### *Anchored vessels and vessels aground*

(a) A vessel at anchor shall exhibit where it can best be seen:

(i) in the fore part, an all-round white light or one ball;

- (ii) at or near the stern and at a lower level than the light prescribed in sub-paragraph (i), an all round white light.

(b) A vessel of less than 50 metres in length may exhibit an all-round white light where it can best be seen instead of the lights prescribed in paragraph (a) of this rule.

(c) A vessel at anchor may, and a vessel of 100 metres and more in length shall, also use the available working or equivalent lights to illuminate her decks.

(d) A vessel aground shall exhibit the lights prescribed in paragraph (a) or (b) of this rule and in addition, where they can best be seen:
- (i) two all-round red lights in a vertical line;
- (ii) three balls in a vertical line.

(e) A vessel of less than 7 metres in length, when at anchor, not in or near a narrow channel, fairway or anchorage, or where other vessels normally navigate, shall not be required to exhibit the lights or shape prescribed in paragraphs (a) and (b) of this rule.

(f) A vessel of less than 12 metres in length, when aground, shall not be required to exhibit the lights or shapes prescribed in sub-paragraphs (d)(i) and (ii) of this rule.

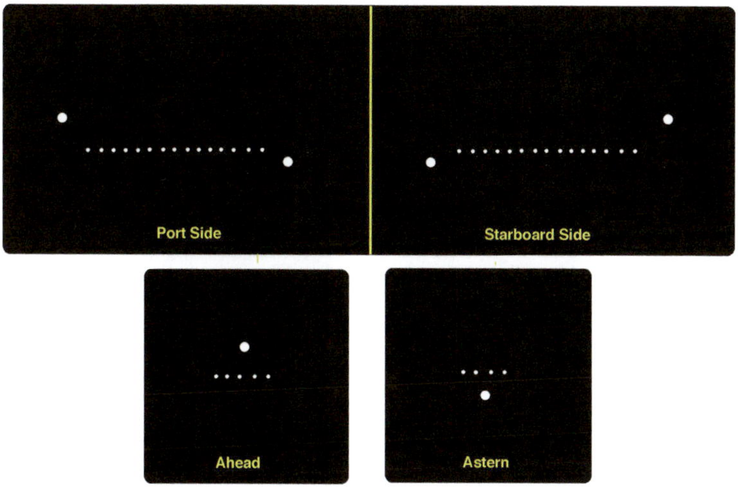


*Figure 13a* A vessel probably 100m or more in length, at anchor

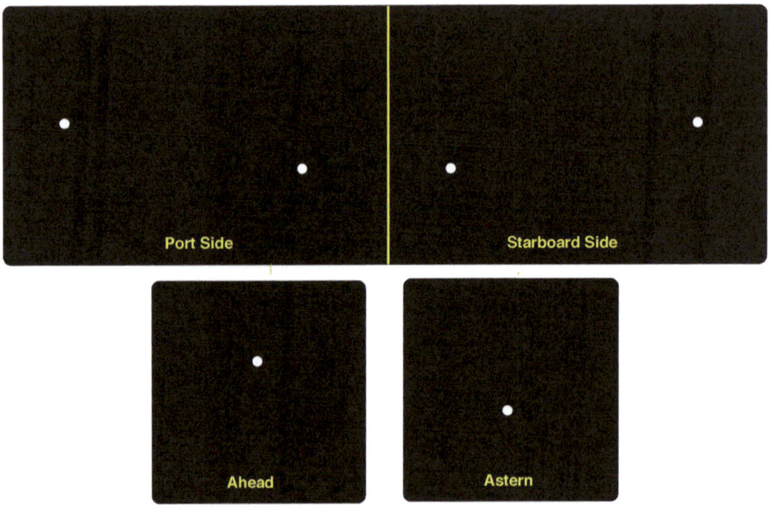


*Figure 13b* A vessel probably 50m or more in length, at anchor

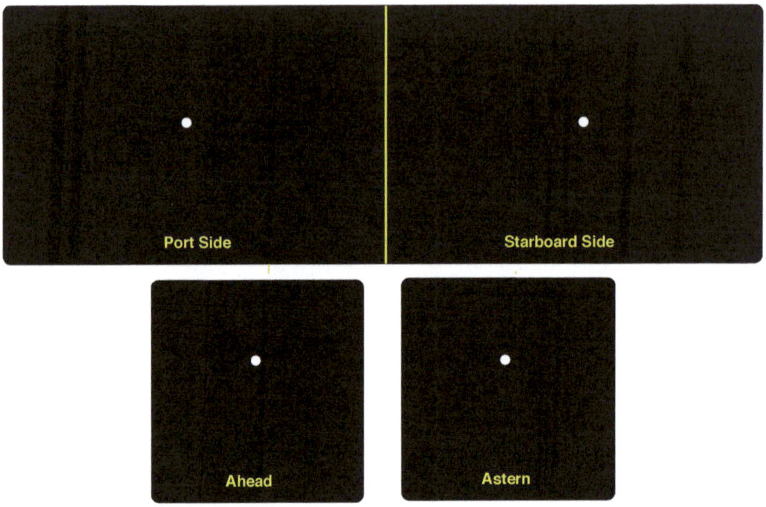


*Figure 13c* A vessel less than 50m in length, at anchor

*Figure 13d* A vessel at anchor, by day, shall exhibit a black ball

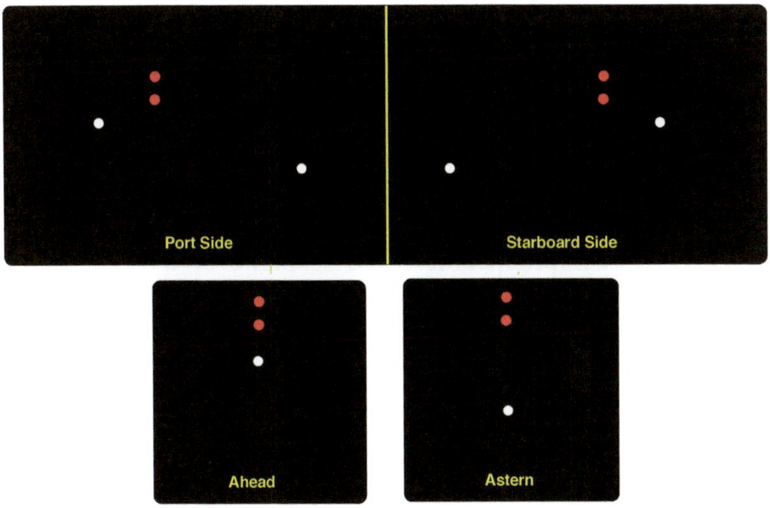


*Figure 13e* A vessel probably 50m or more in length, aground

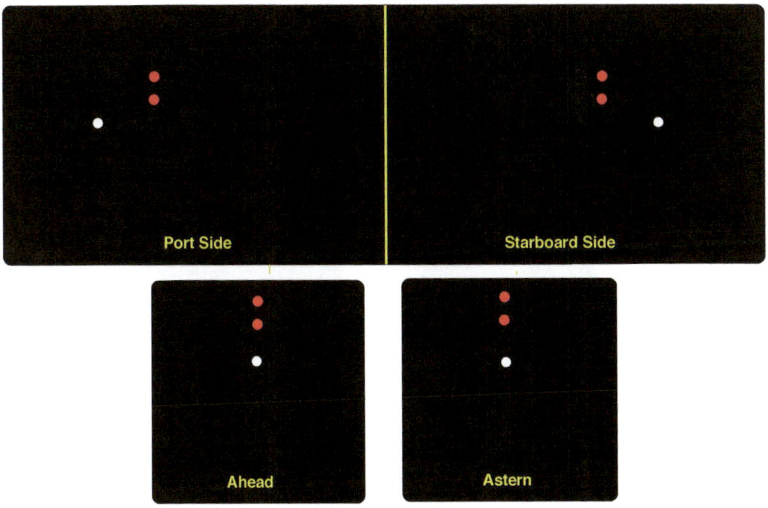


*Figure 13f* A vessel less than 50m in length, aground

*Figure 13g* A vessel aground shall, by day, exhibit three black balls in a vertical line

## Rule 31

### *Seaplanes and WIG craft*

Where it is impracticable for a seaplane or a WIG craft to exhibit lights and shapes of the characteristics or in the positions prescribed in the rules of this part she shall exhibit lights and shapes as closely similar in characteristics and position as is possible.

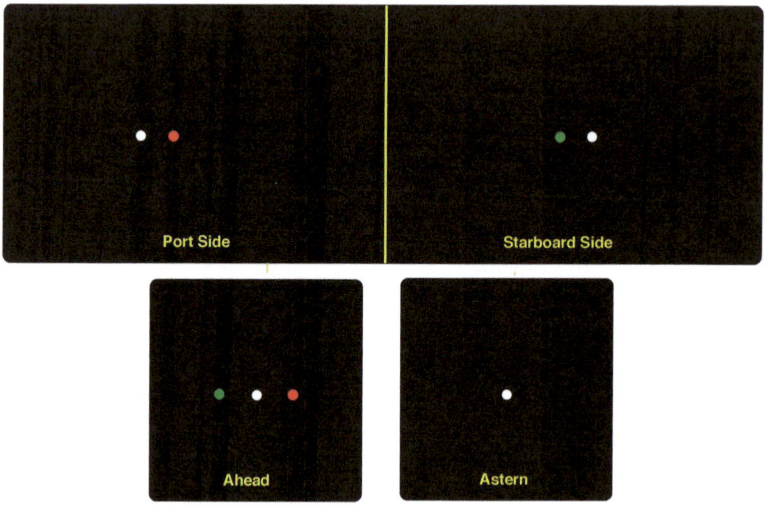


*Figure 14a* Seaplane: light distribution may vary

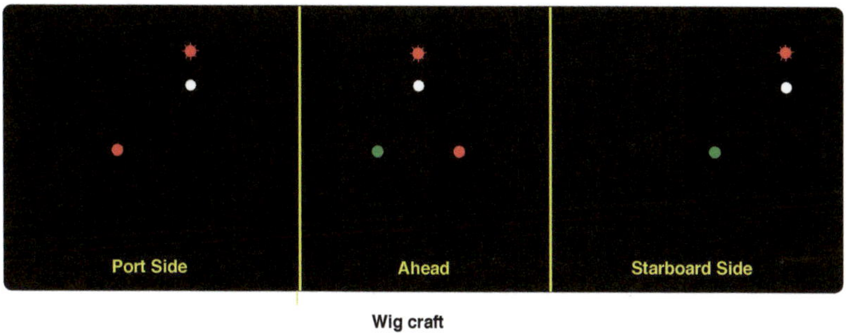


*Figure 14b* WIG craft

## INTERACTIVE NAVIGATION LIGHT IDENTIFICATION QUIZ

(Please visit www.seamanshiptutor.com/rorbook to download the Interactive Navigation Light Identification Quiz.)

## VESSELS UNDERWAY AND VESSELS MAKING WAY OR NOT MAKING WAY

The following table shows which type of vessels are to be considered as 'underway' and which type of vessels as 'making way/not making way'. The right hand column shows whether the length of the vessel can be determined.

*Table C* Vessels underway and vessels making way or not making way

| *Vessel* | *Underway* | *Making way/not making way* | *Can one determine the length?* |
|---|---|---|---|
| Power-driven vessel | x | | Yes |
| Air cushion vessel | x | | Yes |
| Seaplane | x | | Yes |
| WIG craft | x | | Yes |
| Towing vessel | x | | Yes (tug only) |
| Sailing vessel | x | | No |
| Vessel engaged in fishing | | x | No |
| Vessel engaged in trawling | | x | Yes |
| Constrained by her draught | x | | Yes |
| NUC | | x | No |
| RAM – towing | x | | Yes (tug only) |
| RAM – mine clearance | x | | Yes |
| RAM – other | | x | Yes |

# Part D
# Sound and light signals

## Rule 32

### *Definitions*

(a) The word 'whistle' means any sound signalling appliance capable of producing the prescribed blasts and which complies with the specifications in Annex III to these regulations.
(b) The term 'short blast' means a blast of about one second's duration.
(c) The term 'prolonged blasted blast' means a blast of from four to six seconds' duration.

## Rule 33

### *Equipment for sound signals*

(a) A vessel of 12 metres or more in length shall be provided with a whistle, a vessel of 20m or more in length shall be provided with a bell in addition to a whistle, and a vessel of 100m or more in length shall, in addition, be provided with a gong, the tone and sound of which cannot be confused with that of the bell. The whistle, bell and gong shall comply with the specifications in Annex III to these regulations. The bell or gong or both may be replaced by other equipment having the same respective sound characteristics, provided that manual sounding of the prescribed signals shall always be possible.
(b) A vessel of less than 12 metres in length shall not be obliged to carry the sound signalling appliances prescribed in paragraph (a) of this rule but if she does not, she shall be provided with some other means of making an efficient sound signal.

*A Practical Guide to the Rules of the Road, Third edition.* 978-1-138-84389-9.

## Rule 34

### *Manoeuvring and warning signals*

(a) When vessels are in sight of one another, a power-driven vessel underway, when manoeuvring as authorized or required by these rules, shall indicate that manoeuvre by the following signals on her whistle:

- one short blast to mean 'I am altering my course to starboard';
- two short blasts to mean 'I am altering my course to port';
- three short blasts to mean 'I am operating astern propulsion.'

(b) Any vessel may supplement the whistle signals prescribed in paragraph (a) of this rule by light signals, repeated as appropriate, whilst the manoeuvre is being carried out:

(i) these light signals shall have the following significance:

- one flash to mean 'I am altering my course to starboard';
- two flashes to mean 'I am altering my course to port';
- three flashes to mean 'I am operating astern propulsion';

(ii) the duration of each flash shall be about one second, the interval between flashes shall be about one second, and the interval between successive signals shall be not less than ten seconds;

(iii) the light used for this signal shall, if fitted, be an all-round white light, visible at a minimum range of 5 miles, and shall comply with the provisions of Annex I to these regulations.

(c) When in sight of one another in a narrow channel or fairway:

(i) a vessel intending to overtake another shall in compliance with Rule 9(e)(i) indicate her intention by the following signals on her whistle:

- two prolonged blasted blasts followed by one short blast to mean 'I intend to overtake you on your starboard side';
- two prolonged blasted blasts followed by two short blasts to mean 'I intend to overtake you on your port side.'

(ii) the vessel about to be overtaken when acting in accordance with Rule 9(e)(i) shall indicate her agreement by the following signal on her whistle:

- one prolonged blasted, one short blast, one prolonged blasted and one short blast, in that order.

(d) When vessels in sight of one another are approaching each other and from any cause either vessel fails to understand the intentions or actions of the other, or is in doubt whether sufficient action is being taken by the other to avoid collision, the vessel in doubt shall immediately indicate such doubt by giving at least five short and rapid blasts on the whistle. Such signal may be supplemented by a light signal of at least five short and rapid flashes.

(e) A vessel nearing a bend or an area of a channel or fairway where other vessels may be obscured by an intervening obstruction shall sound one prolonged blasted blast. Such signal shall be answered with a prolonged blasted blast by any approaching vessel that may be within hearing around the bend or behind the intervening obstruction.

(f) If whistles are fitted on a vessel at a distance apart of more than 100 metres, one whistle only shall be used for giving manoeuvring and warning signals.

## Rule 35

### *Sound signals in restricted visibility*

In or near an area of restricted visibility, whether by day or night, the signals prescribed in this rule shall be used as follows:

(a) A power-driven vessel making way through the water shall sound at intervals of not more than 2 minutes one prolonged blasted blast.

(b) A power-driven vessel underway but stopped and making no way through the water shall sound at intervals of not more than 2 minutes two prolonged blasted blasts in succession with an interval of about 2 seconds between them.

(c) A vessel not under command, a vessel restricted in her ability to manoeuvre, a vessel constrained by her draught, a sailing vessel, a vessel engaged in fishing and a vessel engaged in towing or pushing another vessel shall, instead of the signals prescribed in paragraphs (a) or (b) of this rule, sound at intervals of not more than 2 minutes three blasts in succession, namely one prolonged blasted blast followed by two short blasts.

(d) A vessel engaged in fishing, when at anchor, and a vessel restricted in her ability to manoeuvre when carrying out her work at anchor, shall instead of the signals prescribed in paragraph (g) of this rule sound the signal prescribed in paragraph (c) of this rule.

(e) A vessel towed or if more than one vessel is towed the last vessel of the tow, if manned, shall at intervals of not more than 2 minutes sound four blasts in succession, namely one prolonged blasted blast followed by three short blasts. When practicable, this signal shall be made immediately after the signal made by the towing vessel.

(f) When a pushing vessel and a vessel being pushed ahead are rigidly connected in a composite unit they shall be regarded as a power-driven vessel and shall give the signals prescribed in paragraphs (a) or (b) of this rule.

(g) A vessel at anchor shall at intervals of not more than one minute ring the bell rapidly for about 5 seconds. In a vessel of 100 metres or more in length the bell shall be sounded in the forepart of the vessel and immediately after the ringing of the bell the gong shall be sounded rapidly for about 5 seconds in the after part of the vessel. A vessel at anchor may in addition sound three blasts in succession, namely one short, one prolonged blasted and one short blast, to give warning of her position and of the possibility of collision to an approaching vessel.

(h) A vessel aground shall give the bell signal and if required the gong signal prescribed in paragraph (g) of this rule and shall, in addition, give three separate and distinct strokes on the bell immediately before and after the rapid ringing of the bell. A vessel aground may in addition sound an appropriate whistle signal.

(i) A vessel of 12 metres or more but less than 20 metres in length shall not be obliged to give the bell signals prescribed in paragraphs (g) and (h) of this rule. However, if she does not, she shall make some other efficient sound signal at intervals of not more than 2 minutes.

(j) A vessel of less than 12 metres in length shall not be obliged to give the above-mentioned signals but, if she does not, shall make some other efficient sound signal at intervals of not more than 2 minutes.

(k) A pilot vessel when engaged on pilotage duty may in addition to the signals prescribed in paragraphs (a), (b) or (g) of this rule sound an identity signal consisting of four short blasts.

## Rule 36

### *Signals to attract attention*

If necessary to attract the attention of another vessel, any vessel may make light or sound signals that cannot be mistaken for any signal authorised elsewhere in these rules, or may direct the beam of her searchlight in the direction of the danger, in such a way as not to embarrass any vessel. Any light to attract the attention of another vessel shall be such that it cannot be mistaken for any aid to navigation. For the purpose of this rule the use of high intensity intermittent or revolving lights, such as strobe lights, shall be avoided.

# Rule 37

## *Distress signals*

**When a vessel is in distress and requires assistance she shall use or exhibit the signals described in Annex IV to these regulations.**

One thing to remember here: 'manoeuvring and warning signals' (Rule 34) used when in sight of one another means in clear visibility only. 'Sound signals in restricted visibility' (Rule 35) will only be used in restricted visibility. Do not get confused between these two types of sound signals, and always use the appropriate signal in the relevant state of visibility.

# Part E
# Exemptions

## Rule 38

### *Exemptions*

Any vessel (or class of vessels) provided that she complies with the requirements of the International Regulations for Preventing Collisions at Sea, 1960(a), the keel of which is laid or which is at a corresponding stage of construction before the entry into force of these regulations may be exempted from compliance therewith as follows:

(a) The installation of lights with ranges prescribed in Rule 22, until 4 years after the date of entry into force of these regulations.

(b) The installation of lights with colour specifications as prescribed in Section 7 of Annex I to these regulations, until 4 years after the date of entry into force of these regulations.

(c) The repositioning of lights as a result of conversion from imperial to metric units and rounding off measurement figures, permanent exemption.

(d) (i) The repositioning of masthead lights on vessels of less than 150 metres in length, resulting from the prescriptions of Section 3(a) of Annex I to these regulations, permanent exemption.

(ii) The repositioning of masthead lights on vessels of 150 metres or more in length, resulting from the prescriptions of Section 3(a) of Annex I to these regulations, until 9 years after the date of entry into force of these regulations.

(e) The repositioning of masthead lights resulting from the prescriptions of Section 2(b) of Annex I to these

*A Practical Guide to the Rules of the Road, Third edition.* 978-1-138-84389-9.

regulations, until 9 years after the date of entry into force of these regulations.

(f) The repositioning of sidelights resulting from the prescriptions of Sections 2(g) and 3(b) of Annex I to these regulations, until 9 years after the date of entry into force of these regulations.

(g) The requirements for sound signal appliances prescribed in Annex III to these regulations, until 9 years after the date of entry into force of these regulations.

(h) The repositioning of all-round lights resulting from the prescription of Section 9(b) of Annex I to these regulations, permanent exemption.

# Open sea situations

## SITUATION 01

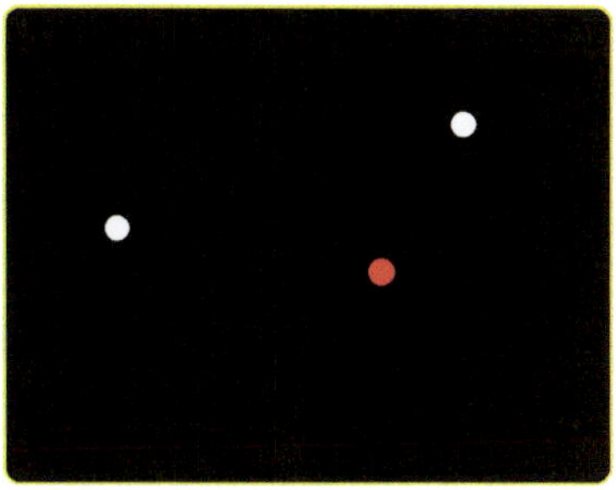

**Q.** You are the OOW of a power-driven vessel and you see these lights. What do the lights indicate to you?

**A.** A power-driven vessel probably 50m or more in length, port side on, underway.

**Q.** You see this vessel four points on the starboard bow. What action will you take?

**A.** I will take a series of compass bearings of an approaching vessel and, if the compass bearing does not change appreciably, then the risk of collision exists.

**Q.** The bearing is constant and range decreasing. What action will you take now?

**A.** As per Rule 15, my vessel is the give-way vessel and the other is to stand on. I will sound one short blast and make a broad alteration of the course to starboard and pass well clear of the other vessel.

## SITUATION 02

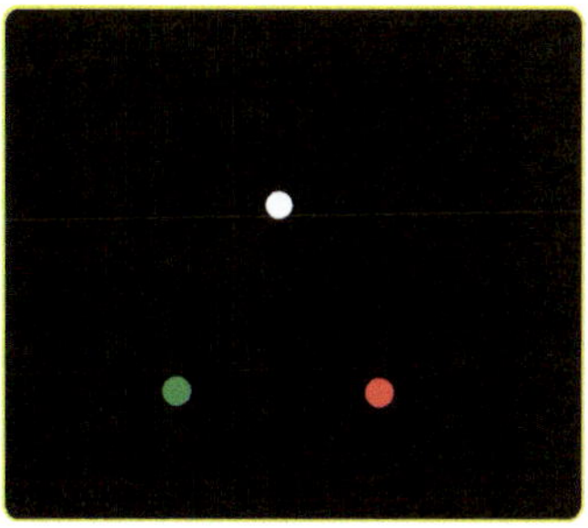

**Q.** You are the OOW of a power-driven vessel. You see these lights. What do the lights indicate to you?

**A.** A power-driven vessel less than 50m in length, underway, end-on.

*A Practical Guide to the Rules of the Road, Third edition.* 978-1-138-84389-9.

**Q.** You see this vessel right ahead. What action will you take?

**A.** As per Rule 14, my vessel is the give-way vessel and the other is also the give-way vessel. I will sound one short blast and then make a broad alteration of the course to starboard and pass well clear of the other vessel.

## SITUATION 03

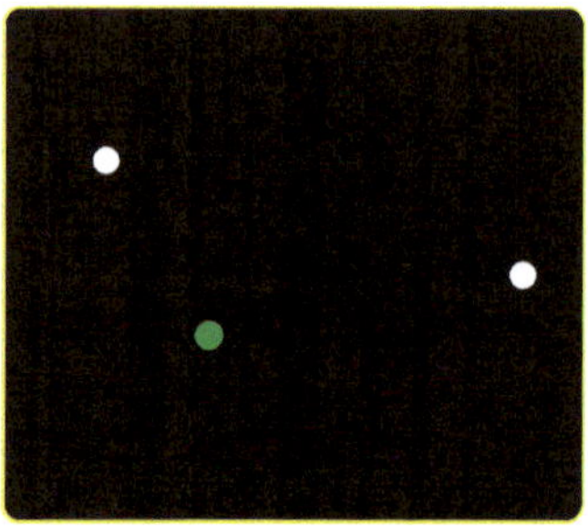

**Q.** You are the OOW of a power-driven vessel. You see these lights. What do the lights indicate to you?

**A.** A power-driven vessel probably 50m or more in length, underway, seen from the starboard side.

**Q.** You see this vessel three points on the port bow. What action will you take?

**A.** I will take a series of compass bearings of an approaching vessel and, if the compass bearing does not change appreciably, then the risk of collision exists.

**Q.** The bearing is constant and range decreasing. What action will you take now?

**A.** As per Rule 15 my vessel is the stand-on vessel and other vessel is the give-way vessel. I will maintain my course and speed and stand on with caution and continue to monitor the situation.

**Q.** This vessel is not taking any action and getting closer. What will you do now?

**A.** I will sound at least five short and rapid blasts to draw her attention. At night I will also use an Aldis lamp to draw her attention.

**Q.** She is still not taking any action and getting even closer. What will be your action now?

**A.** As per Rule 17, since it is apparent to me that the other vessel is not taking any action, I will take the action by sounding one short blast and alter course to starboard to take a round turn and pass astern of her. As per Rule 17 I will not alter course to port for a vessel on my own port side.

## SITUATION 04

**Q.** You are the OOW of a power-driven vessel. You see this light. What do the light indicate to you?

**A.** A sailing vessel underway, port side on, not showing optional sailing lights.

**Q.** You see this vessel 30° on the starboard bow. What action will you take?

**A.** I will take a series of compass bearings of the approaching vessel and, if the compass bearing does not change appreciably, then the risk of collision exists.

**Q.** The bearing is constant and range decreasing. What action will you take now?

**A.** As per Rule 18 it is my responsibility to keep clear of a sailing vessel so my vessel is the give-way vessel and the sailing vessel is to stand on. I will sound one short blast and make a broad alteration of the course to starboard and pass well clear of the other vessel.

## SITUATION 05

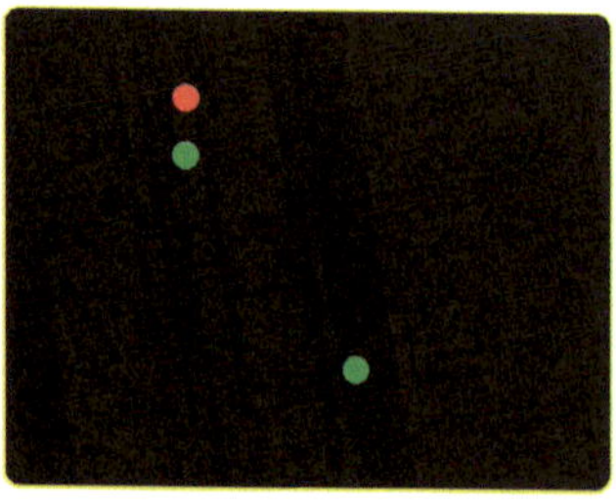

**Q.** You are the OOW of a power-driven vessel. You see these lights. What do the lights indicate to you?

**A.** A sailing vessel underway, starboard side on, showing optional sailing lights.

**Q.** You see this vessel 40° on the port bow. What action will you take?

**A.** I will take a series of compass bearings of an approaching vessel and, if the compass bearing does not change appreciably, then the risk of collision exists.

**Q.** The bearing is constant and range decreasing. What action will you take now?

**A.** As per Rule 18 it is my responsibility to keep clear of a sailing vessel so my vessel is the give-way vessel and the sailing vessel is to stand on. I will sound two short blasts on the whistle and make a broad alteration of the course to port and pass round her stern.

## SITUATION 06

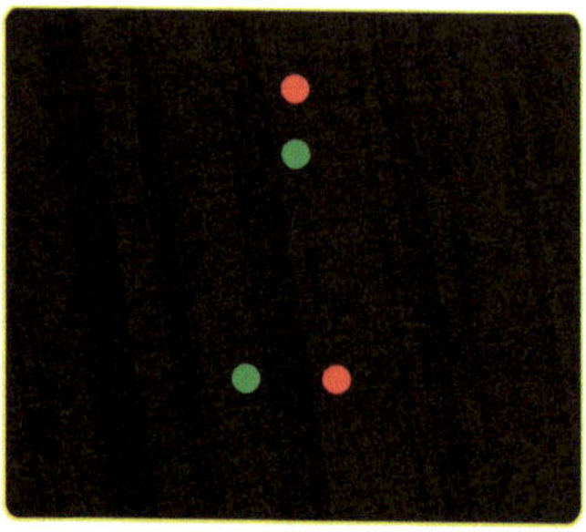

**Q.** You are the OOW of a power-driven vessel. You see these lights. What do the lights indicate to you?

**A.** A sailing vessel underway, end-on, showing optional sailing lights.

**Q.** You see this vessel right ahead. What action will you take?

**A.** I will deem risk of collision to exist. As per Rule 18 it is my responsibility to keep clear of a sailing vessel so my vessel is the give-way vessel and the sailing vessel is to stand on. I will sound one short blast on the whistle and make a broad alteration of the course to starboard.

**Q.** Does Rule 14 apply here?
**A.** No. Rule 14 is between two power-driven vessels.
**Q.** Why did you alter course to starboard?
**A.** This is the ordinary practice of seamen.
**Q.** Could you have altered course to port?
**A.** Yes I could have, but I prefer the starboard side.
**Q.** Would you consider wind direction when taking action for a sailing vessel?
**A.** If I am on a power-driven vessel I would not consider wind direction. I will keep clear of her.

## SITUATION 07

**Q.** You are the OOW of a power-driven vessel. You see this light. What it could be?
**A.** It could be one of following:

1 A power-driven vessel underway seen from astern.
2 A sailing vessel underway seen from astern.
3 A vessel of less than 50m in length at anchor.
4 A power-driven vessel of less than 7m in length with a maximum speed not exceeding 7 knots.
5 A sailing vessel underway less than 7 metres in length.
6 A vessel under oars.
7 A life raft.

**Q.** This is a power-driven vessel seen from astern. You see this light one point on the starboard bow. What action will you take?
**A.** I will take a series of compass bearings of an approaching vessel and, if the compass bearing does not change appreciably, then the risk of collision exists.
**Q.** The bearing is constant and range decreasing. What action will you take now?

**A.** As per Rule 13, I am an overtaking vessel in this situation and it is my responsibility to keep clear of the vessel being overtaken. I will sound one short blast on the whistle and alter course to starboard.

**Q.** Can you alter course to port?

**A.** Yes I can. Rule 13 does not specify which way to alter course. The rule only says to keep clear of the vessel being overtaken. I would prefer to alter course to starboard and cross her stern and keep my starboard side clear.

**Q.** Some other night, you see a vessel's side light then the stern light and then the same side light again, two points on your port bow. What would you assume?

**A.** As per Rule 13 if I am in doubt then I will assume I am an overtaking the vessel and shall keep clear until finally past and clear.

## SITUATION 08

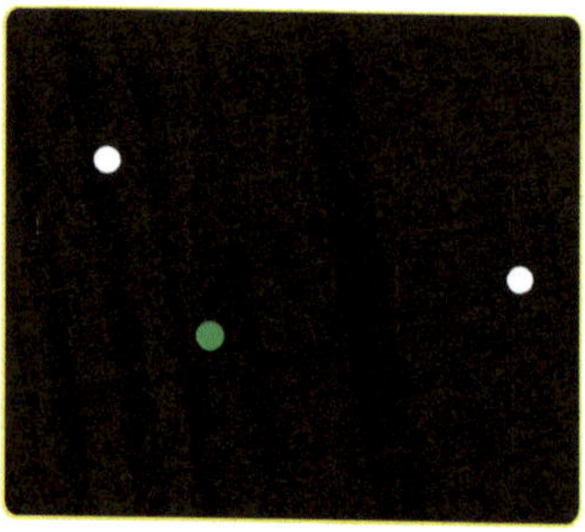

**Q.** You are the OOW of a power-driven vessel. You see these lights. What do the lights indicate to you?

**A.** A power-driven vessel probably 50m or more in length, underway, seen from the starboard side.

**Q.** You see this vessel four points abaft the port beam. What action will you take?

**A.** I will take a series of compass bearings of an approaching vessel and, if the compass bearing does not change appreciably, then the risk of collision exists.

**Q.** The bearing is constant and range decreasing. What action will you take now?

**A.** As per Rule 13, my vessel is the stand-on vessel and the other vessel is to give way. I will maintain my course and speed and stand on with caution and continue to monitor the situation.

**Q.** This vessel is not taking any action and getting close. What will you do now?
**A.** I will sound at least five short and rapid blasts to draw her attention. At night I will also use the Aldis lamp to draw her attention.
**Q.** She is still not taking any action and getting close. What will be your action now?
**A.** As per Rule 17, since it is apparent to me that the other vessel is not taking any action, I will take the action by sounding one short blast and alter course to starboard and take a round turn and pass astern of the other vessel.

## SITUATION 09

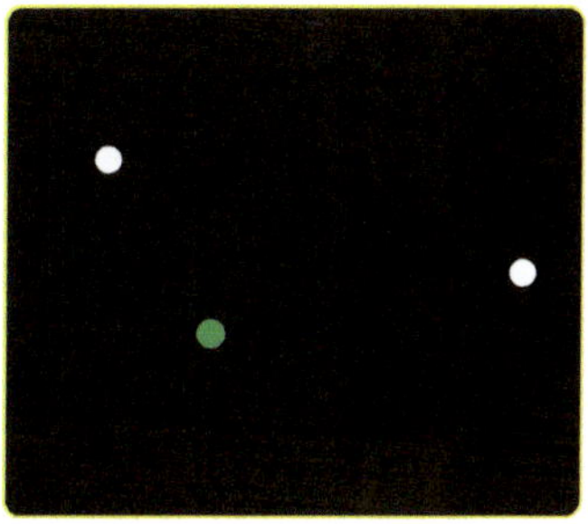

**Q.** You are the OOW of a power-driven vessel. You see these lights. What do the lights indicate to you?
**A.** A power-driven vessel probably 50m or more in length, underway, seen from the starboard side.
**Q.** You see this vessel on the port quarter. What action will you take?
**A.** I will take a series of compass bearings of an approaching vessel and, if the compass bearing does not change appreciably, then the risk of collision exists.
**Q.** The bearing is constant and range decreasing. What action will you take now?
**A.** As per Rule 13, my vessel is the stand-on vessel and the other vessel is to give way. I will maintain my course and speed and stand on with caution and continue to monitor the situation.
**Q.** The other vessel takes action and overtakes from your starboard beam, and then alters course to her port and becomes a crossing vessel. What will you do now?

**A.** As per Rule 13(d), an overtaking vessel shall not become a crossing vessel with any subsequent alteration of the bearing. It is the duty of the overtaking vessel to keep clear of the overtaken vessel until she is finally past and clear. I will stand on with caution and monitor the situation.

## SITUATION 10

**Q.** You are the OOW of a power-driven vessel. You see these lights. What do the lights indicate to you?

**A.** A power-driven vessel engaged in towing, less than 50m in length, length of the tow does not exceed 200m, seen from port side, underway.

**Q.** You see this vessel 40° on the starboard bow. What action will you take?

**A.** I will take a series of compass bearings of an approaching vessel and, if the compass bearing does not change appreciably, then the risk of collision exists.

**Q.** How will you take the bearings of this vessel?

**A.** I will take the bearing of both, the tug and the tow lights.

**Q.** The bearing changes slightly and the range is decreasing. What action will you take now?

**A.** The risk of collision may exist even if the bearing is changing when approaching a tow. As per Rule 15, my vessel is the give-way vessel and the other is to stand on. I will sound one short blast and make a broad alteration of the course to starboard and pass round the stern of the vessel being towed.

## SITUATION 11

**Q.** You are the OOW of a power-driven vessel. You see these lights. What do the lights indicate to you?

**A.** A power-driven vessel engaged in towing, less than 50m in length, length of tow more than 200m, starboard side on, underway.

**Q.** You see this vessel 30° on the port bow. What action will you take?

**A.** I will take a series of compass bearings of an approaching vessel and, if the compass bearing does not change appreciably, then the risk of collision exists.

**Q.** The bearing is constant and range decreasing. What action will you take now?

**A.** As per Rule 15, my vessel is the stand-on vessel and the other vessel is to give way. I will maintain my course and speed and stand on with caution and continue to monitor the situation.

**Q.** But this is a towing vessel. Don't you think that this is a stand-on vessel?

**A.** The towing vessel and her tow are not severely restricted to deviate from their course so I will treat the target as normal power-driven vessel.

## SITUATION 12

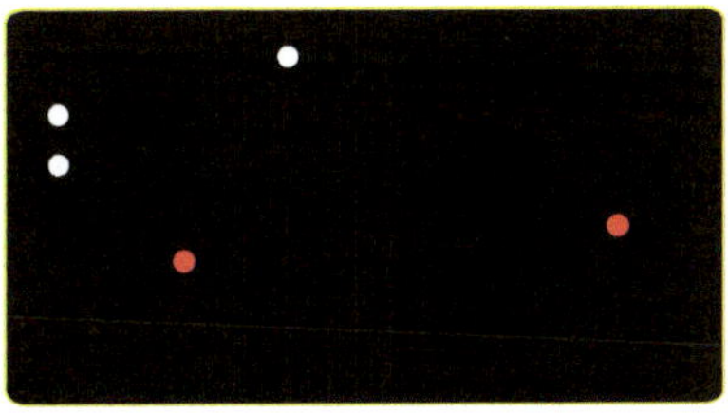

**Q.** You are the OOW of a power-driven vessel. You see these lights. What do the lights indicate to you?

**A.** A power-driven vessel engaged in towing, probably 50m or more in length, length of tow less than 200m, port side on, underway.

**Q.** You see this vessel 35° on the starboard bow. What action will you take?

**A.** I will take a series of compass bearings of an approaching vessel and if the bearing does not change appreciably then the risk of collision exists.

**Q.** The bearing changes slightly and the range is decreasing. What action will you take now?

**A.** As per Rule 15, my vessel is the give-way vessel and the other is to stand on. I will sound one short blast and make a broad alteration of the course to starboard and pass round the stern of the vessel being towed.

**Q.** What fog signal shall be sounded by the towing vessel?

**A.** The towing vessel shall at intervals of not more than 2 minutes sound three blasts in succession, namely one prolonged blasted blast followed by two short blasts.

**Q.** What fog signal shall be sounded by the vessel being towed?

**A.** The vessel being towed, if manned, shall at intervals of not more than 2 minutes sound four blasts in succession, namely one prolonged blasted blast followed by three short blasts. When practicable, this signal shall be made immediately after the signal made by the towing vessel.

## SITUATION 13

**Q.** You are the OOW of a power-driven vessel. You see these lights. What do the lights indicate to you?

**A.** A power-driven vessel engaged in towing, seen from astern, underway.

**Q.** Can you tell me the length of the tow?

**A.** No. The information on the length of tow can only be obtained from the masthead lights.

**Q.** You see this vessel 15° on the starboard bow. What action will you take?

**A.** I will take a series of compass bearings of an approaching vessel and if the compass bearing does not change appreciably then the risk of collision exists.

**Q.** The bearing changes slightly with range decreasing. What action will you take now?

**A.** As per Rule 13, my vessel is the give-way vessel and the other is to stand on. I will sound one short blast and make a broad alteration of the course to starboard and pass round the stern of the vessel being towed.

**Q.** Can you alter course to port?

**A.** Yes I can but I prefer to alter course to starboard to keep my starboard side clear. Rule 13 requires you to keep clear of the vessel being overtaken. It does not give the action to take, nor what not to do.

## SITUATION 14

**Q.** You are the OOW of a power-driven vessel. You see these lights. What do the lights indicate to you?

**A.** A power-driven vessel engaged in towing, probably 50m or more in length, length of tow more than 200m, tug and tow severely restricted from being able to deviate from their course, seen from the starboard side, underway.

**Q.** You see this vessel 45° on the port bow. What action will you take?

**A.** I will take a series of compass bearings of an approaching vessel and if the compass bearing does not change appreciably then the risk of collision exists.

**Q.** The bearing changes slightly and the range is decreasing. What action will you take now?
**A.** As per Rule 18, my vessel is the give-way vessel and the other is to stand on. I will sound two short blasts and make a broad alteration of the course to port and pass round the stern of the vessel being towed.
**Q.** What fog signal shall be sounded by the towing vessel?
**A.** The towing vessel shall sound at intervals of not more than 2 minutes three blasts in succession, namely one prolonged blasted blast followed by two short blasts.
**Q.** What is the fog signal of a vessel being towed?
**A.** A vessel being towed, if manned, shall at intervals of not more than 2 minutes sound four blasts in succession, namely one prolonged blasted blast followed by three short blasts. When practicable, this signal shall be made immediately after the signal made by the towing vessel.

## SITUATION 15

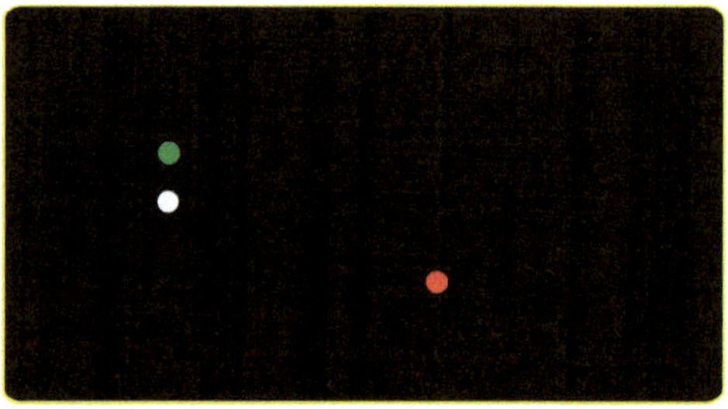

**Q.** You are the OOW of a power-driven vessel. You see these lights. What do the lights indicate to you?
**A.** A vessel engaged in trawling, less than 50m in length, seen from the port side, making way through the water.
**Q.** You see this vessel 30° on the starboard bow. What action will you take?
**A.** I will take a series of compass bearings of an approaching vessel and if the compass bearing of the approaching vessel does not change appreciably then the risk of collision exists.

**Q.** The bearing is constant and range decreasing. What action will you take now?

**A.** As per Rule 18, my vessel is the give-way vessel and the other is to stand on. I will sound one short blast and make a broad alteration of the course to starboard and pass round the stern of the vessel engaged in trawling.

**Q.** What is the fog signal of the vessel engaged in trawling?

**A.** A vessel engaged in trawling shall sound at intervals of not more than 2 minutes three blasts in succession, namely one prolonged blasted blast followed by two short blasts.

## SITUATION 16

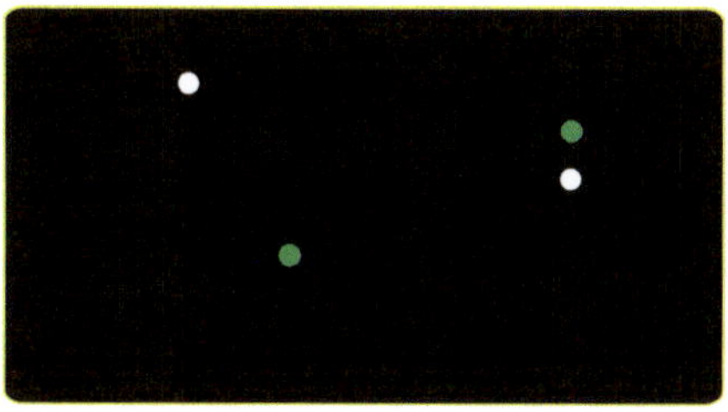

**Q.** You are the OOW of a power-driven vessel. You see these lights. What do the lights indicate to you?

**A.** A vessel engaged in trawling, probably 50m or more in length, starboard side on, making way through the water.

**Q.** You see this vessel 40° on the port bow. What action will you take?

**A.** I will take a series of compass bearings of an approaching vessel and, if the compass bearing does not change appreciably, then the risk of collision exists.

**Q.** The bearing is constant and range decreasing. What action will you take now?

**A.** As per Rule 18, my vessel is the give-way vessel and the other is to stand on. I will sound two short blasts and make a broad alteration of the course to port and pass round the stern of the vessel engaged in trawling.

**Q.** What is the fog signal of the vessel engaged in trawling?

**A.** A vessel engaged in trawling shall sound at intervals of not more than 2 minutes three blasts in succession, namely one prolonged blasted blast followed by two short blasts.

## SITUATION 17

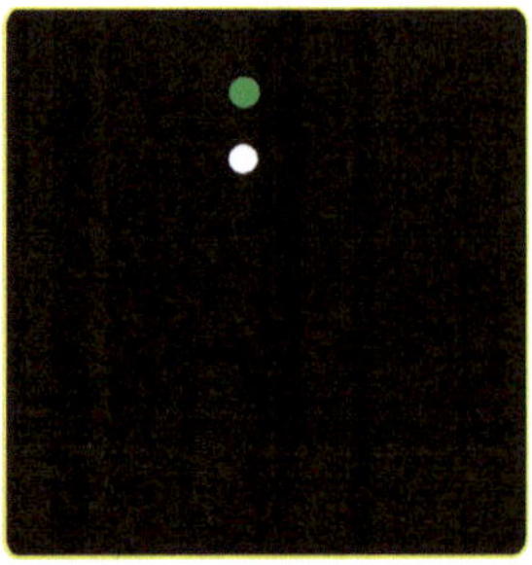

**Q.** You are the OOW of a power-driven vessel. You see these lights. What do the lights indicate to you?

**A.** A vessel engaged in trawling, less than 50m in length, not making way through the water or at anchor.

**Q.** You see this vessel right ahead. What action will you take?

**A.** In this case risk of collision exists. I will sound one short blast and make a broad alteration of the course to starboard and pass clear of the vessel engaged in trawling.

**Q.** What is the fog signal of the vessel engaged in trawling?

**A.** A vessel engaged in trawling shall sound at intervals of not more than 2 minutes three blasts in succession, namely one prolonged blasted blast followed by two short blasts.

## SITUATION 18

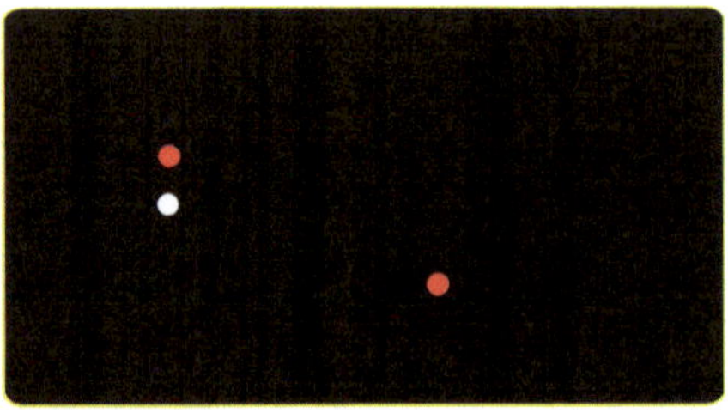

**Q.** You are the OOW of a power-driven vessel. You see these lights. What do the lights indicate to you?

**A.** A vessel engaged in fishing, other than trawling, port side on, making way through the water.

**Q.** You see this vessel 30° on the starboard bow. What action will you take?

**A.** I will take a series of compass bearings of an approaching vessel and if the compass bearing does not change appreciably then the risk of collision exists.

**Q.** The bearing is constant and range decreasing. What action will you take now?

**A.** As per Rule 18, my vessel is the give-way vessel and the other is to stand on. I will sound one short blast and make a broad alteration of the course to starboard and pass round the stern of the vessel engaged in fishing.

**Q.** What is the fog signal of the vessel engaged in fishing?

**A.** A vessel engaged in fishing shall sound at intervals of not more than 2 minutes three blasts in succession, namely one prolonged blast followed by two short blasts.

## SITUATION 19

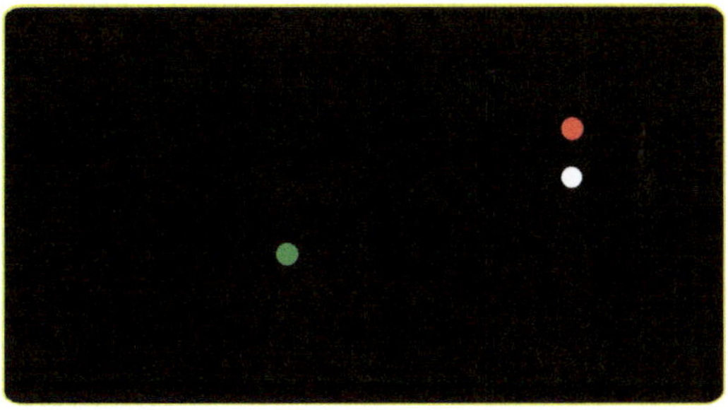

**Q.** You are the OOW of a power-driven vessel. You see these lights. What do the lights indicate to you?

**A.** A vessel engaged in fishing, other than trawling, seen from the starboard side, making way through the water.

**Q.** You see this vessel 20° on the port bow. What action will you take?

**A.** I will take a series of compass bearings of an approaching vessel and if the compass bearing does not change appreciably then the risk of collision exists.

**Q.** The bearing is constant and range decreasing. What action will you take now?

**A.** As per Rule 18, my vessel is the give-way vessel and the vessel engaged in fishing is to stand on. I will sound two short blasts and then make a broad alteration of the course to port and pass round the stern of the vessel engaged in fishing.

## SITUATION 20

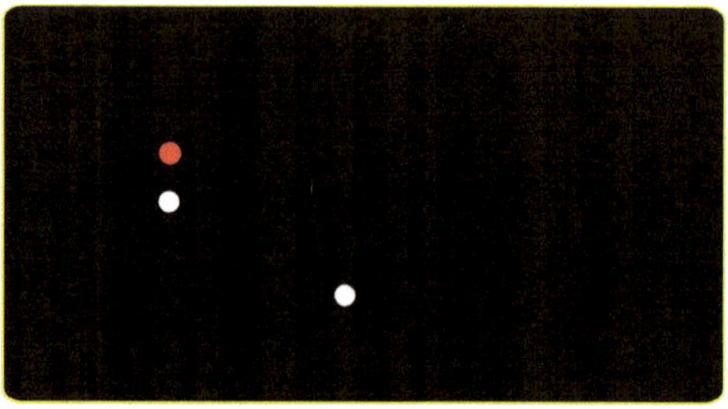

**Q.** You are the OOW of a power-driven vessel. You see these lights. What do the lights indicate to you?

**A.** A vessel engaged in fishing, other than trawling, underway but stopped and making no way through the water; OR at anchor, with outlying gear extending more than 150 metres horizontally from the vessel in the direction of the lower white light.

**Q.** Do you know the length of the nets?

**A.** No.

**Q.** You see this vessel 20° on the port bow. What action will you take?

**A.** As per Rule 18, my vessel is the give-way vessel and the vessel engaged in fishing is to stand on. I will sound two short blasts and then make a broad alteration of the course to port to pass round the red and white vertical lights.

**Q.** Can you cross over the nets?

**A.** I shall avoid crossing over the nets.

## SITUATION 21

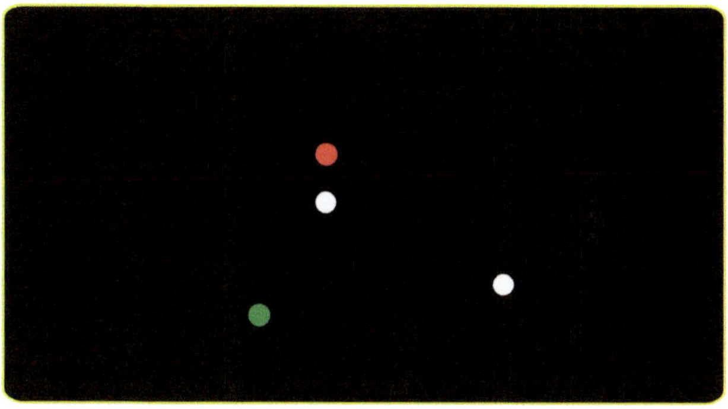

**Q.** You are the OOW of a power-driven vessel. You see these lights. What do the lights indicate to you?

**A.** A vessel engaged in fishing, other than trawling, starboard side on, making way through the water, with outlying gear extending more than 150 meters horizontally from the vessel in the direction of the lower white light.

**Q.** You see this vessel 25° on the port bow. What action will you take?

**A.** I will take a series of compass bearings of an approaching vessel and if the bearing does not change appreciably then the risk of collision exists.

**Q.** The bearing is constant and range decreasing. What action will you take now?

**A.** As per Rule 18, my vessel is the give-way vessel and the vessel engaged in fishing is to stand on. I will sound two short blasts and then make a broad alteration of the course to port to pass clear of the nets.

**Q.** Can you cross the nets?

**A.** I shall avoid crossing over the nets.

**Q.** How would you cross the nets if it became necessary?

**A.** I would stop engines and make sure the propeller is stopped; and then cross over the nets with the help of head reach.

## SITUATION 22

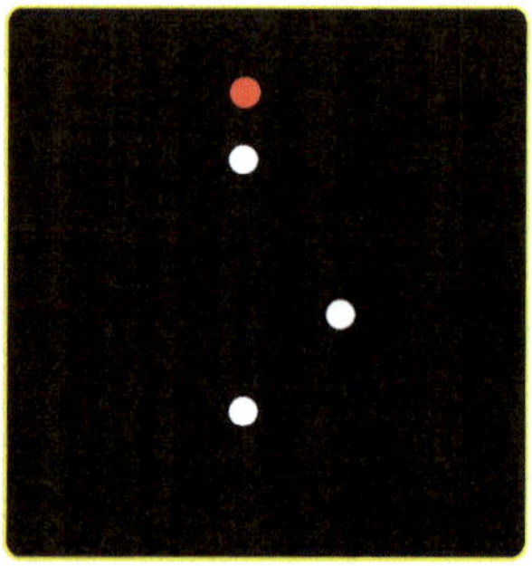

**Q.** You are the OOW of a power-driven vessel. You see these lights. What do the lights indicate to you?

**A.** A vessel engaged in fishing, other than trawling, seen from astern, making way through the water, with outlying gear extending more than 150 meters horizontally from the vessel in the direction of the white light on her starboard side.

**Q.** You see this vessel 15° on the port bow. What action will you take?

**A.** Since I am an overtaking vessel, as per Rule 13 my vessel is the give-way vessel and the vessel engaged in fishing is to stand on. I will sound two short blasts and then make a broad alteration of the course to port to pass round it.

**Q.** Can you alter course to starboard to overtake her?

**A.** I shall avoid crossing over her nets.

**Q.** What is the fog signal of the vessel engaged in fishing?

**A.** A vessel engaged in fishing shall sound at intervals of not more than 2 minutes three blasts in succession, namely one prolonged blast followed by two short blasts.

## SITUATION 23

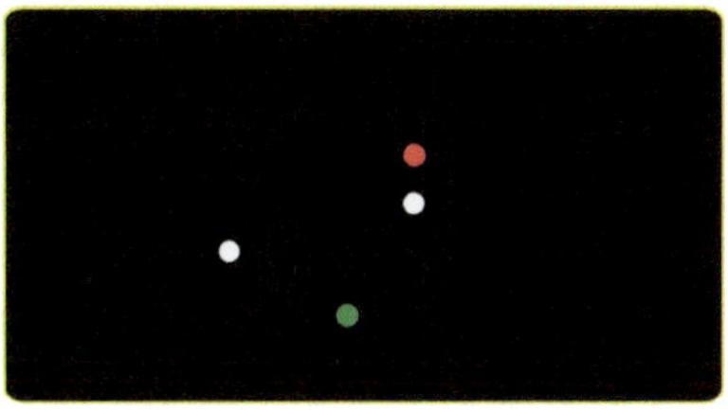

**Q.** You are the OOW of a power-driven vessel. You see these lights. What do the lights indicate to you?

**A.** A vessel engaged in fishing, other than trawling, seen from the starboard side, making way through the water, with outlying gear extending more than 150 meters horizontally from the vessel in the direction of the lower white light.

**Q.** You see this vessel 20° on the port bow. What action will you take?

**A.** I will take a series of compass bearings of an approaching vessel and if the bearing does not change appreciably then the risk of collision exists.

**Q.** The bearing is constant and range decreasing. What action will you take now?

**A.** As per Rule 18, my vessel is the give-way vessel and the vessel engaged in fishing is to stand on. I will sound one short blast and then make a broad alteration of the course to starboard and steer parallel to her course; and then once I have overtaken her I will pass well ahead.

**Q.** Can you alter course to port?

**A.** I shall avoid altering course to port because of her nets.

## SITUATION 24

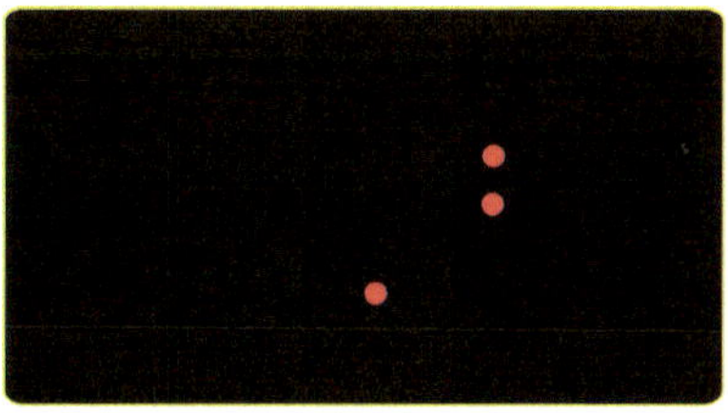

**Q.** You are the OOW of a power-driven vessel. You see these lights. What do the lights indicate to you?

**A.** A vessel not under command, port side on, making way through the water.

**Q.** What is a vessel not under command?

**A.** The term 'vessel not under command' means a vessel which through some exceptional circumstance is unable to manoeuvre as required by these rules and is therefore unable to keep out of the way of another vessel.

**Q.** You see this vessel 30° on the starboard bow. What action will you take?

**A.** I will take a series of compass bearings of an approaching vessel and, if the compass bearing does not change appreciably, then the risk of collision exists.

**Q.** The bearing is constant and range decreasing. What action will you take now?

**A.** As per Rule 18, my vessel is the give-way vessel and the vessel not under command is to stand on. I will sound one short blast and make a broad alteration of the course to starboard and pass round the stern of the vessel not under command.

**Q.** What is the fog signal of the vessel not under command?

**A.** A vessel not under command will sound at intervals of not more than 2 minutes three blasts in succession, namely one prolonged blast followed by two short blasts.

## SITUATION 25

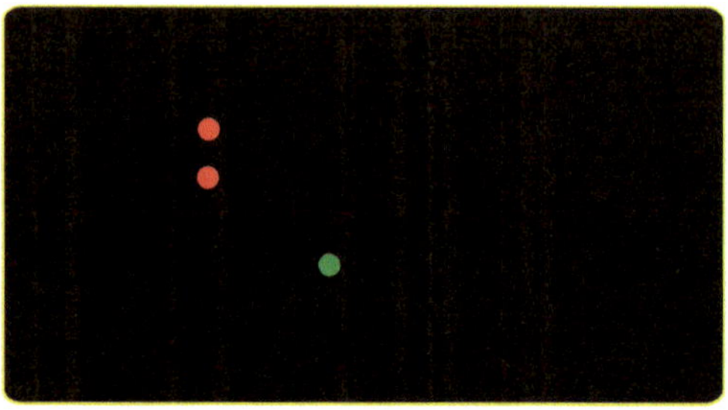

**Q.** You are the OOW of a power-driven vessel. You see these lights. What do the lights indicate to you?

**A.** A vessel not under command, starboard side on, making way through the water.

**Q.** You see this vessel 35° on the port bow. What action will you take?

**A.** I will take a series of compass bearings of an approaching vessel, and if the bearing does not change appreciably then the risk of collision exists.

**Q.** The bearing is constant and range decreasing. What action will you take now?

**A.** As per Rule 18, my vessel is the give-way vessel and the vessel not under command is to stand on. I will sound two short blasts and then make a broad alteration of the course to port side and pass round the stern of the vessel not under command.

**Q.** In which situation does the vessel not under command become the give-way vessel?

**A.** In the highly unlikely event of a vessel not under command becoming an overtaking vessel, then as per Rule 13 she becomes the give-way vessel.

## SITUATION 26

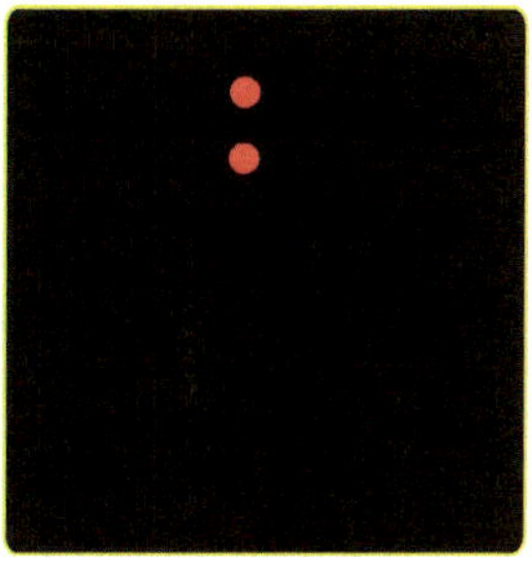

**Q.** You are the OOW of a power-driven vessel. You see these lights. What do the lights indicate to you?

**A.** A vessel not under command, underway and not making way.

**Q.** You see this vessel right ahead. What action will you take?

**A.** I will deem a risk of collision to exist and, as per Rule 18, my vessel is the give-way vessel and the vessel not under command is to stand on. I will sound one short blast and then make a broad alteration of the course to starboard side and pass clear of the vessel not under command.

**Q.** What is the day signal of the vessel not under command?

**A.** Two black balls in a vertical line where they can best be seen.

## SITUATION 27

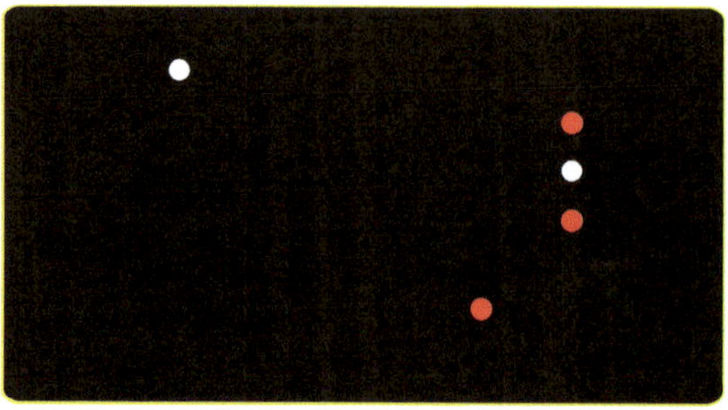

**Q.** You are the OOW of a power-driven vessel. You see these lights. What do the lights indicate to you?

**A.** A vessel restricted in her ability to manoeuvre, less than 50m in length, seen from the port side, making way through the water.

**Q.** What do you mean by 'a vessel restricted in her ability to manoeuvre'?

**A.** The term 'a vessel restricted in her ability to manoeuvre' means a vessel which by the nature of her work is restricted in her ability to manoeuvre as required by these rules and is therefore unable to keep out of the way of another vessel. The term 'vessels restricted in their ability to manoeuvre' shall include but not be limited to:

- (i) a vessel engaged in laying, servicing or picking up a navigation mark, submarine cable or pipeline;
- (ii) a vessel engaged in dredging, surveying or underwater operations;
- (iii) a vessel engaged in replenishment or transferring persons, provisions or cargo while underway;
- (iv) a vessel engaged in the launching or recovery of aircraft;
- (v) a vessel engaged in mine-clearance operations;
- (vi) a vessel engaged in a towing operation such as severely restricts the towing vessel and her tow in their ability to deviate from their course.

**Q.** You see this vessel 30° on the starboard bow. What action will you take?

**A.** I will take a series of compass bearings of an approaching vessel and, if the compass bearing does not change appreciably, then the risk of collision exists.

**Q.** The bearing is constant and range decreasing. What action will you take now?

**A.** As per Rule 18, my vessel is the give-way vessel and the vessel restricted in her ability to manoeuvre is to stand on. I will sound one short blast and make a broad alteration of the course to starboard and pass round the stern of the vessel restricted in her ability to manoeuvre.

## SITUATION 28

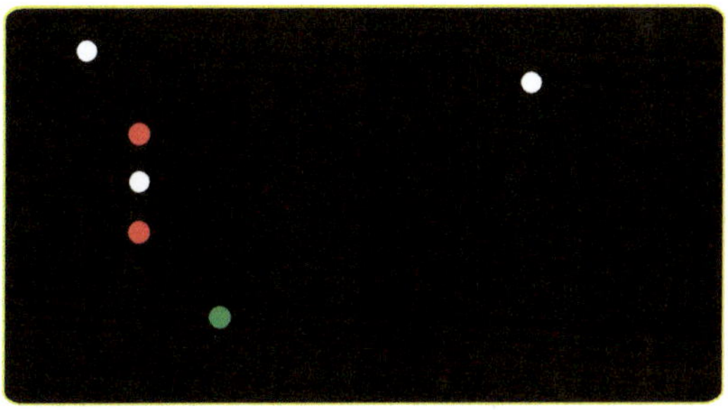

**Q.** You are the OOW of a power-driven vessel. You see these lights. What do the lights indicate to you?

**A.** A vessel restricted in her ability to manoeuvre, probably 50m or more in length, seen from the starboard side, making way through the water.

**Q.** You see this vessel 40° on the port bow. What action will you take?

**A.** I will take a series of compass bearings of an approaching vessel and, if the compass bearing does not change appreciably, then the risk of collision exists.

**Q.** The bearing is constant and range decreasing. What action will you take now?

**A.** As per Rule 18, my vessel is the give-way vessel and the vessel restricted in her ability to manoeuvre is to stand on. I will sound two short blasts and then make a broad alteration of the course to port side and pass round the stern of the vessel restricted in her ability to manoeuvre.

**Q.** What is the fog signal of the vessel restricted in her ability to manoeuvre?

**A.** A vessel restricted in her ability to manoeuvre shall sound at intervals of not more than 2 minutes three blasts in succession, namely one prolonged blast followed by two short blasts.

## SITUATION 29

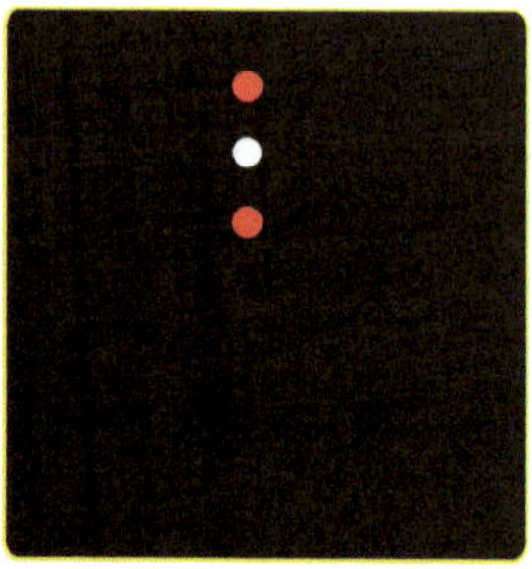

**Q.** You are the OOW of a power-driven vessel. You see these lights. What do the lights indicate to you?

**A.** A vessel restricted in her ability to manoeuvre, underway and not making way through water.

**Q.** You see this vessel right ahead. What action will you take?

**A** I will deem a risk of collision to exist and as per Rule 18, my vessel is the give-way vessel and the vessel restricted in her ability to manoeuvre is to stand on. I will sound one short blast and then make a broad alteration of the course to starboard to pass clear of the vessel restricted in her ability to manoeuvre.

**Q.** What is the day signal of the vessel restricted in her ability to manoeuvre?

**A.** Three shapes in a vertical line where they can best be seen. The highest and lowest of these shapes shall be black balls and the middle one a black diamond.

**Q.** Can these shapes be of different colours?

**A.** As per Annex I these shapes shall be black in colour.

## SITUATION 30

**Q.** You are the OOW of a power-driven vessel. You see these lights. What do the lights indicate to you?

**A.** A vessel restricted in her ability to manoeuvre, engaged in dredging or underwater operations, probably 50m or more in length, end on, making way through the water.

**Q.** You see this vessel right ahead. What action will you take?

**A** I will deem a risk of collision to exist and as per Rule 18, my vessel is the give-way vessel and the vessel restricted in her ability to manoeuvre is to stand on. I will sound two short blasts and make a broad alteration of the course to port and pass clear of the vessel restricted in her ability to manoeuvre.

**Q.** Can you alter course to starboard?

**A.** No. The two red lights in a vertical line indicate that the vessel has an obstructed side, and that I must keep clear of her port side in this case. The two green lights identify the clear side – which is starboard side of the vessel in this case.

**Q.** How shall she indicate the clear side and obstructed side by day?

**A.** Two black balls in a vertical line indicate the obstructed side; and two black diamonds in a vertical line indicate the clear side – where other vessels may pass.

**Q.** What is the fog signal of the vessel engaged in dredging operation?

**A.** A vessel engaged in dredging operations, when restricted in her ability to manoeuvre, shall sound at intervals of not more than 2 minutes three blasts in succession, namely one prolonged blast followed by two short blasts.

## SITUATION 31

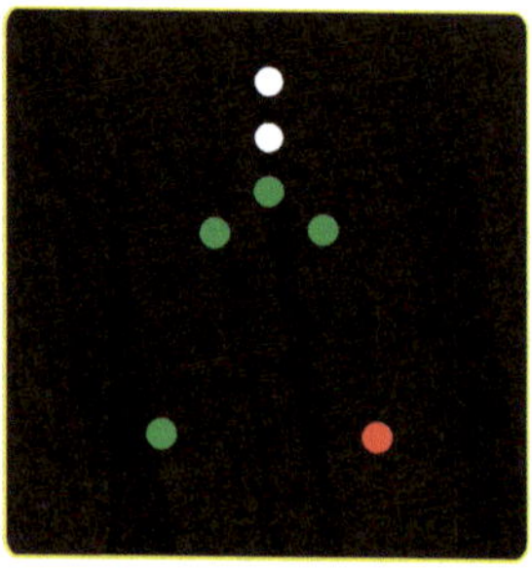

**Q.** You are the OOW of a power-driven vessel. You see these lights. What do the lights indicate to you?

**A.** A vessel engaged in mine-clearance operations, probably 50m or more in length, end-on, underway.

**Q.** You see this vessel right ahead. What action will you take?

**A** I will call the master [not applicable if you are appearing for the Master's Examination] and establish communication with the vessel engaged in mine-clearance operations. I will deem a risk of collision to exist in this situation and as per Rule 18, my vessel is the give-way vessel and the vessel restricted in her ability to manoeuvre, engaged in mine-clearance operations, is to stand on. I will sound one short blast and then make a broad alteration of the course to starboard, passing at least 1000m clear of the vessel engaged in mine-clearance operations.

**Q.** What is the day signal of the vessel engaged in mine-clearance operations?

**A.** She shall exhibit three black balls. One ball shall be exhibited near the foremast head and one at each end of the fore yard.

**Q.** What is the fog signal of the vessel engaged in mine-clearance operations?

**A.** A vessel engaged in mine-clearance operations shall sound at intervals of not more than 2 minutes three blasts in succession, namely one prolonged blast followed by two short blasts.

## SITUATION 32

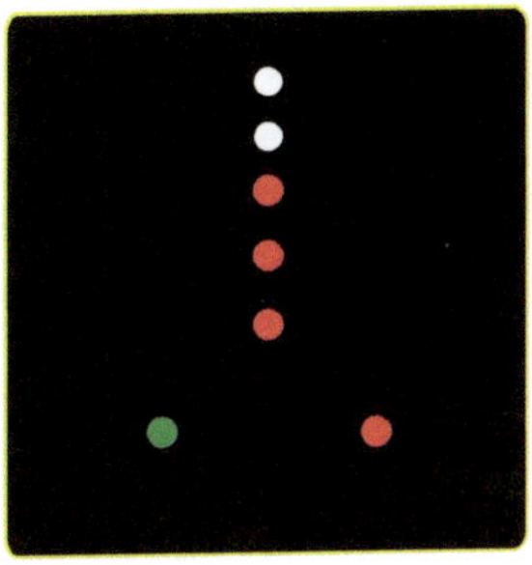

**Q.** You are the OOW of a power-driven vessel. You see these lights. What do the lights indicate to you?

**A.** A vessel constrained by her draught, probably 50m or more in length, end on, underway.

**Q.** You see this vessel right ahead. What action will you take?

**A** I will deem a risk of collision to exist in this situation and as per Rule 18, I shall avoid impeding the safe passage of the vessel constrained by her draught so I will sound one short blast and make a broad alteration of the course to starboard, passing clear of the vessel constrained by her draught.

**Q.** What do you mean by 'vessel constrained by her draught'?

**A.** The term 'vessel constrained by her draught' means a power-driven vessel which, because of her draught in relation to the available depth and width of navigable water, is severely restricted in her ability to deviate from the course she is following.

**Q.** What is the day signal of the vessel constrained by her draught?

**A.** She shall exhibit a black cylinder.

**Q.** What is the fog signal of the vessel constrained by her draught?

**A.** A vessel constrained by her draught shall sound at intervals of not more than 2 minutes three blasts in succession, namely one prolonged blast followed by two short blasts.

## SITUATION 33

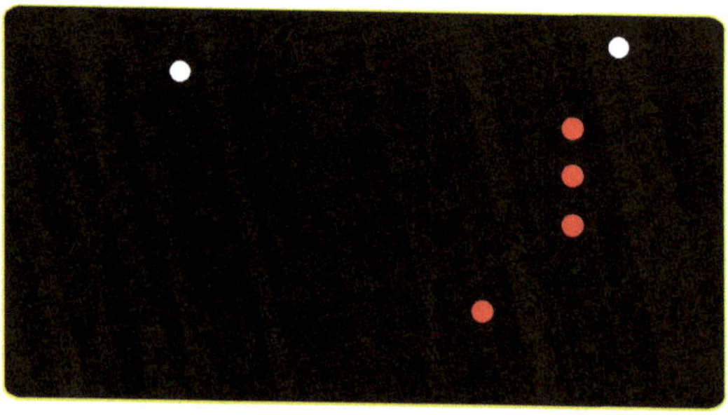

**Q.** You are the OOW of a power-driven vessel. You see these lights. What do the lights indicate to you?

**A.** A vessel constrained by her draught, probably 50m or more in length, port side on, underway.

**Q.** You see this vessel 30° on the starboard bow. What action will you take?

**A.** I will take a series of compass bearings of an approaching vessel and, if the compass bearing does not change appreciably, then the risk of collision exists.

**Q.** The bearing is constant and range decreasing. What action will you take now?

**A** As per Rule 18, I shall avoid impeding the safe passage of the vessel constrained by her draught so I will sound one short blast and then make a broad alteration of course to starboard and pass round the stern of the vessel constrained by her draught.

**Q.** What is the fog signal of the vessel constrained by her draught?

**A.** A vessel constrained by her draught shall sound at intervals of not more than 2 minutes three blasts in succession, namely one prolonged blast followed by two short blasts.

## SITUATION 34

**Q.** You are the OOW of a power-driven vessel. You see these lights. What do the lights indicate to you?

**A.** A power-driven vessel constrained by her draught, probably 50m or more in length, seen from the starboard side, underway.

**Q.** You see this vessel 45° on the port bow. What action will you take?

**A.** I will take a series of compass bearings of an approaching vessel and, if the compass bearing does not change appreciably, then the risk of collision exists.

**Q.** The bearing is constant and range decreasing. What action will you take now?

**A** As per Rule 18, I shall avoid impeding the safe passage of the vessel constrained by her draught, so I will reduce my speed in ample time and let the vessel constrained by her draught pass ahead of me.

**Q.** Why would you reduce speed for this vessel? Can you not alter course to port and pass astern of her?

**A.** As per Rule 18, a vessel constrained by her draught is not a privileged vessel like 'a vessel not under command' or 'a vessel engaged in fishing'. As per Rule 8(f)(ii), when taking action to provide safe passage to a vessel constrained by her draught I have to take action according to the rules of this part and since I am the stand-on vessel and the power-driven vessel constrained by her draught is the give-way vessel, and as per Rule 17(c), I shall not alter course to port for a vessel on my own port side. As per Rule 8(f)(iii), a vessel constrained by her draught remains fully obliged to comply with the rules of this part when the two vessels are approaching one another so as to involve risk of collision.

**Q.** Can you alter course to your starboard side?

**A.** Yes. If altering course to starboard I will take a round turn and pass astern of the vessel constrained by her draught.

## SITUATION 35

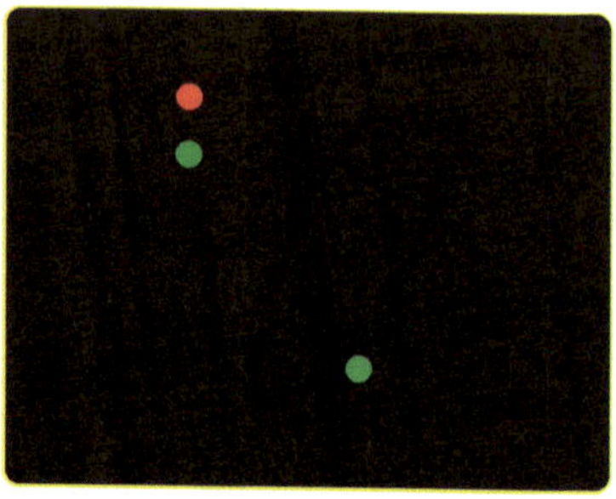

**Q.** You are the OOW of a power-driven vessel constrained by her draught in coastal waters. You see these lights. What do the lights indicate to you?

**A.** A sailing vessel underway, starboard side on, showing optional sailing lights.

**Q.** You see this vessel 40° on the port bow. What action will you take?

**A.** I will take a series of compass bearings of an approaching vessel and, if the compass bearing does not change appreciably, then the risk of collision exists.

**Q.** The bearing is constant and range decreasing. What action will you take now?

**A.** As per Rule 18(d), any vessel other than a vessel not under command or a vessel restricted in her ability to manoeuvre shall, if the circumstances of the case admit, avoid impeding the safe passage of a vessel constrained by her draught. So, I would expect the sailing vessel to take action to avoid collision.

**Q.** What if she does not take action?

**A.** Then I will reduce my speed and let the sailing vessel pass ahead of my vessel.

## SITUATION 36

**Q.** You are the OOW of a power-driven vessel. You see these lights. What do the lights indicate to you?

**A.** A vessel engaged on pilotage duties, underway, starboard side on.

**Q.** You see this vessel three points on the port bow. What action will you take?

**A.** I will take a series of compass bearings of an approaching vessel and, if the compass bearing does not change appreciably, then the risk of collision exists.

**Q.** The bearing is constant and range decreasing. What action will you take now?

**A.** Assuming this is a power-driven vessel, as per Rule 15, my vessel is the stand-on vessel and the other vessel is the give-way vessel. I will maintain my course and speed and stand on with caution, continuing to monitor the situation.

**Q.** Why do you assume it is a power-driven vessel?

**A.** Rule 29(a) states 'a vessel engaged on pilotage duty', which means it could be a power-driven vessel, or a vessel under oars, or a sailing vessel not showing optional lights.

## SITUATION 37

**Q.** You are the OOW of a power-driven vessel. You see these lights. What do the lights indicate to you?

**A.** A vessel engaged on pilotage duties, probably 50m or more in length, at anchor.

## SITUATION 38

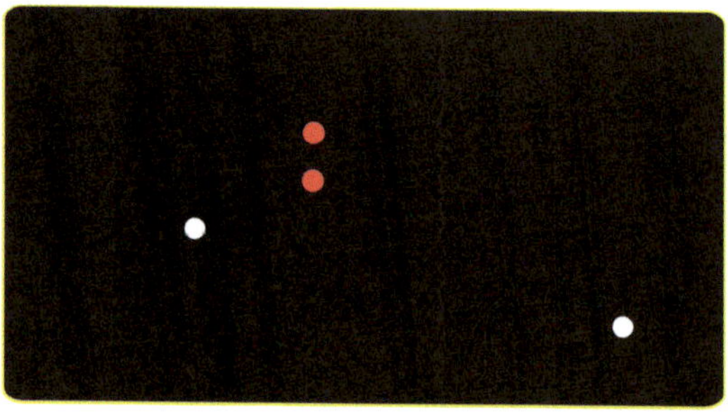

**Q.** You are the OOW of a power-driven vessel. You see these lights. What do the lights indicate to you?

**A.** A vessel probably 50m or more in length, aground.

**Q.** You see this vessel right ahead. What action will you take?

**A** I will stop my vessel and take all way off. I will call the master, do a chart assessment and establish communication with the vessel.

## SITUATION 39

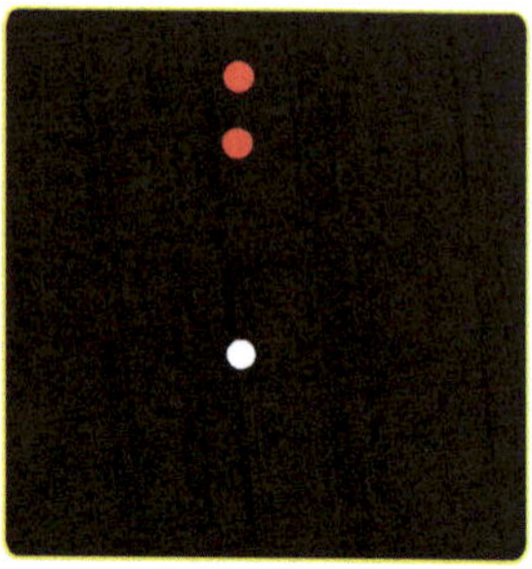

**Q.** You are the OOW of a power-driven vessel. You see these lights. What do the lights indicate to you?

**A.** It could be a vessel less than 50m, aground; **OR** a vessel not under command, seen from astern, underway. In deep waters it can only be a vessel not under command, seen from astern, underway.

## SITUATION 40

**Q.** You are the OOW of a power-driven vessel. You see these lights. What do the lights indicate to you?

**A.** An air cushion vessel or non-displacement craft less than 50m in length, underway, seen from the starboard side.

**Q.** You see this vessel four points on the port bow. What action will you take?

**A.** I will take a series of compass bearings of an approaching vessel and, if the compass bearing does not change appreciably, then the risk of collision exists.

**Q.** The bearing is constant and range decreasing. What action will you take now?

**A.** As per Rule 15 my vessel is the stand-on vessel and other vessel is to give way. I will maintain my course and speed and stand on with caution and continue to monitor the situation.

## SITUATION 41

**Q.** You are the OOW of a power-driven vessel. You see these lights. What do the lights indicate to you?

**A.** A wing-in-ground craft operating in WIG mode, less than 50m in length, underway, seen from the starboard side.

**Q.** What do you mean by operating in WIG mode?

**A.** A WIG craft will display a red flashing light only when taking off, landing and in flight near the surface.

**Q.** What are the characteristics of the red flashing light?
**A.** A high-intensity all-round flashing red light, at 120 flashes per minute.
**Q.** You see this vessel three points on the port bow. What action will you take?
**A.** I will take a series of compass bearings of an approaching vessel and, if the compass bearing does not change appreciably, then the risk of collision exists.
**Q.** The bearing is constant and range decreasing. What action will you take now?
**A.** As per Rule 18(f), a WIG craft, when taking off, landing and in flight near the surface, shall keep well clear of the other vessels and avoid impeding their navigation. So I would expect her to take action to avoid collision.
**Q.** It is apparent to you that she is not taking action. What will you do?
**A.** I will sound one short blast and alter course to starboard and go parallel with her course and re-assess the situation.

## SITUATION 42

**Q.** You are the OOW of a power-driven vessel. You see these lights. What do the lights indicate to you?
**A.** A power-driven vessel engaged in towing, less than 50m in length, tow length of more than 200m, starboard side on, underway.

**Q.** You see this vessel four points on the port bow. What action will you take?

**A.** I will take a series of compass bearings of an approaching vessel and, if the compass bearing does not change appreciably, then the risk of collision exists.

**Q.** The bearing is constant and range decreasing. What action will you take now?

**A.** As per Rule 15, my vessel is the stand-on vessel and the other vessel is to give way. I will maintain my course and speed and stand on with caution and continue to monitor the situation.

**Q.** This vessel is not taking any action and getting closer. What will you do now?

**A.** I will sound at least five short and rapid blasts to draw her attention. At night I will also use the Aldis lamp to draw her attention.

**Q.** She is still not taking any action and getting even closer. What will be your action now?

**A.** As per Rule 17(a)(ii), since it is apparent to me that other vessel is not taking any action I will take the action by sounding one short blast then alter course to starboard and go parallel with her course and re-assess the situation.

**Q.** Can you go to port?

**A.** As per Rule 17(c) I will not alter course to port for a power-driven vessel on my own port side.

## SITUATION 43

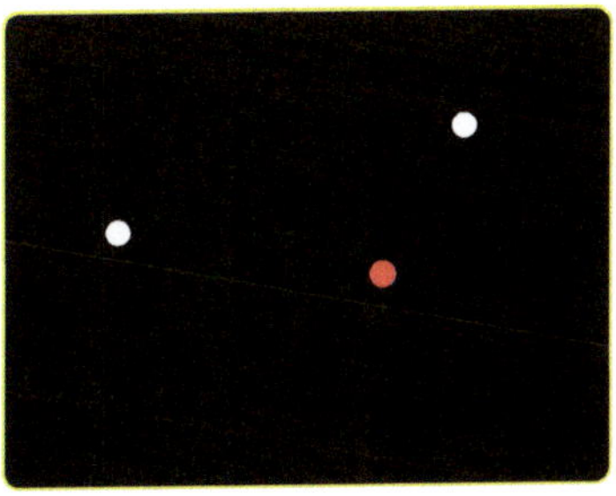

**Q.** You are the OOW of a power-driven vessel and you see these lights. What do the lights indicate to you?

**A.** A power-driven vessel probably 50m or more in length, port side on, underway.

**Q.** You see this vessel four points on the starboard bow. What action will you take?

**A.** I will take a series of compass bearings of an approaching vessel and, if the compass bearing does not change appreciably, then the risk of collision exists.

**Q.** The bearing is constant and range decreasing. What action will you take now?

**A.** As per Rule 15, my vessel is the give-way vessel and the other is to stand on. I will sound one short blast and make a broad alteration of the course to starboard and pass well clear of the other vessel.

**Q.** There is shallow patch on your starboard which means you cannot alter course to starboard. What can you do now?

**A.** I will reduce my speed in ample time.

**Q.** Is there anything else you can do?

**A.** I can alter course to port and steer parallel to the other vessel's course and then re-assess the situation. Rule 15 only says 'avoid crossing ahead of the other vessel' and by altering course to port and steering parallel to other vessel's course my action will be within the rules.

## SITUATION 44

**Q.** You are the OOW of a power-driven vessel. You are involved in assisting another vessel in distress. A decision has been made that your vessel will tow the vessel in distress. The length of your vessel is more than 50m and the length of the tow is more than 200m. What navigational lights will you display as a towing vessel?

**A.** If available, I will display three masthead lights in a vertical line in the forward and one masthead light abaft of and higher than the forward masthead lights; sidelights; a sternlight; and a towing light in a vertical line above the sternlight.

As per Rule 24(i), where from sufficient cause it is impracticable for a vessel not normally engaged in towing operations to display the lights prescribed above, such a vessel shall not be required to exhibit those lights when engaged in towing another vessel in distress. I will make sure to indicate the relationship between the towing vessel and the vessel being towed by illuminating the towline.

**Q.** Can you use a high-intensity, intermittent, or revolving light to illuminate the towline?

**A.** As per Rule 36, I shall avoid use of a high-intensity, intermittent, or revolving light to illuminate the towline.

# TSS and narrow channel situations

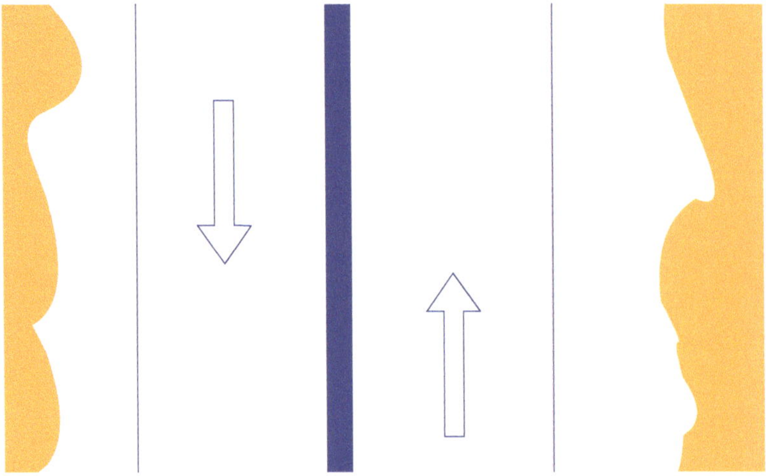

## SITUATION 01

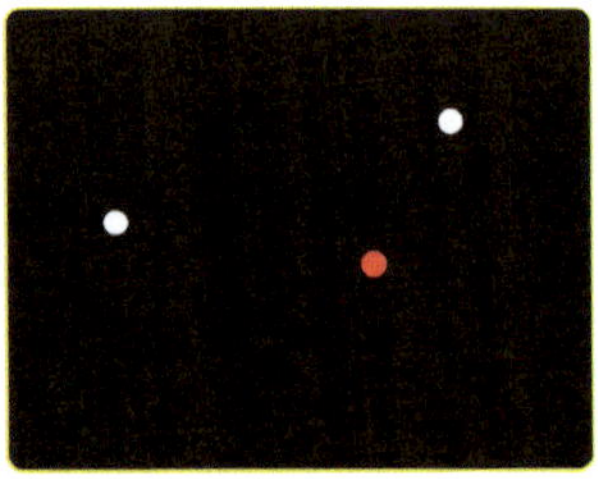

**Q.** You are the OOW of a power-driven vessel. You are in a Traffic Separation Scheme (TSS) and proceeding in an appropriate traffic lane in the general direction of traffic flow for that lane. You see these lights. What do the lights indicate to you?

**A.** A power-driven vessel probably 50m or more in length, port side on, underway.

**Q.** You see this vessel four points on the starboard bow. What action will you take?

**A.** I will take a series of compass bearings of an approaching vessel and, if the compass bearing does not change appreciably, then the risk of collision exists.

**Q.** The bearing is constant and range decreasing. What action will you take now?

**A.** As per Rule 15, my vessel is the give-way vessel and the other is to stand on. In this situation I will make sure there is enough room behind my ship and reduce speed to let the other vessel pass ahead of me.

**Q.** There is a vessel right behind you. You cannot reduce your speed.

**A.** In that case I will sound one short blast and make a broad alteration of the course to starboard to pass well clear of the other vessel and, if necessary, go to the inshore traffic zone to avoid collision.

**Q.** But you are in a traffic separation scheme. Can you go into an inshore traffic zone?

**A.** Rule 10 does not relieve any vessel of her obligation under any other rule. So as to avoid collision with a vessel on my starboard side, I shall alter course to starboard and go into the inshore traffic zone.

## SITUATION 02

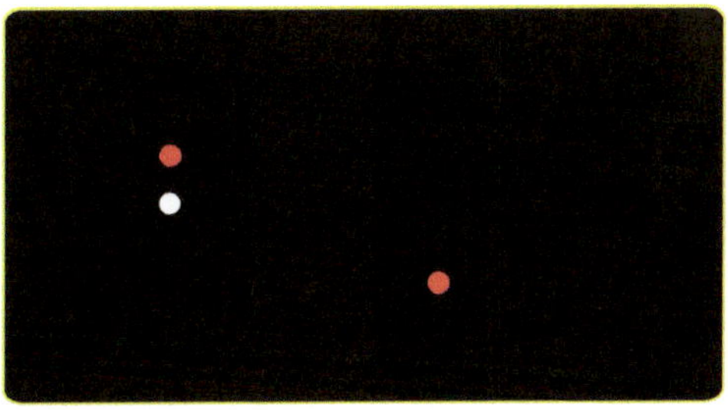

**Q.** You are the OOW of a power-driven vessel. You are in a traffic separation scheme and proceeding in an appropriate traffic lane in the general direction of traffic flow for that lane. You see these lights. What do the lights indicate to you?

**A.** It is a vessel engaged in fishing, port side on, making way through the water.

**Q.** You see this vessel 30° on the starboard bow. What action will you take?

**A.** I will take a series of compass bearings of an approaching vessel and, if the compass bearing does not change appreciably, then the risk of collision exists.

**Q.** The bearing is constant and range decreasing. What action will you take now?

**A.** As per Rule 10(i), a vessel engaged in fishing shall not impede the passage of any vessel following a traffic lane. I would expect the vessel engaged in fishing to take action to avoid collision.

**Q.** She is not taking any action and getting closer. What would you do now?

**A.** I will sound at least five short and rapid blasts on the whistle to draw her attention. At night I will also use the Aldis lamp to draw her attention. I will also inform Vessel Traffic Services (VTS).

**Q.** She is still not taking any action and getting even closer. What will be your action now?

**A.** I will sound one short blast and then make a broad alteration of the course to starboard and pass well clear of the other vessel and go into the inshore traffic zone to avoid collision.

## SITUATION 03

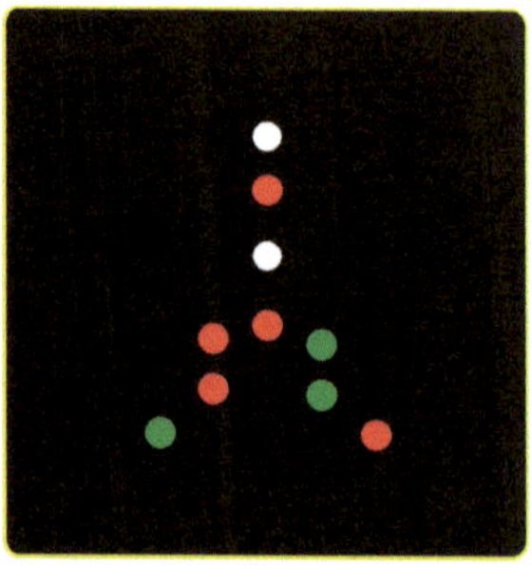

**Q.** You are the OOW of a power-driven vessel. You are in a traffic separation scheme and proceeding in an appropriate traffic lane in the general direction of traffic flow for that lane. You see these lights. What do the lights indicate to you?

**A.** It is a vessel engaged in dredging operations, less than 50m in length, restricted in her ability to manoeuvre, end-on, making way through the water.

**Q.** You see this vessel right ahead. What action will you take?

**A.** As per Rule 10(k), a vessel restricted in her ability to manoeuvre when engaged in an operation for the maintenance of safety of navigation in a traffic separation scheme is exempted from complying with this rule to the extent necessary to carry out the operation. I will make a slight alteration of course to starboard to pass on her port side which is the clear side in this case.

**Q.** Why do you make a slight alteration?

**A.** To keep my vessel nearly in the general direction of the traffic flow and in the lane.

## SITUATION 04

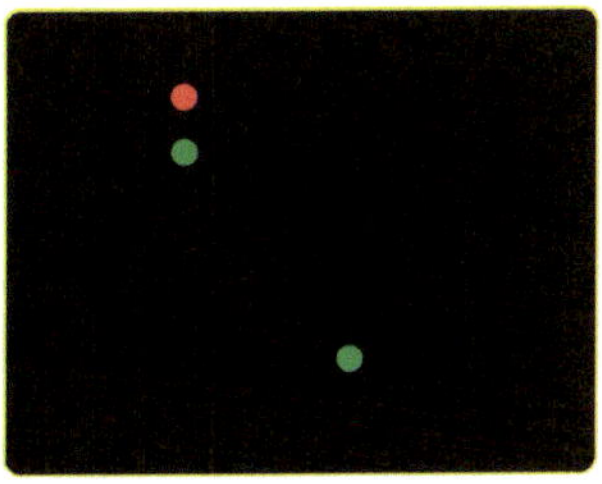

**Q.** You are the OOW of a power-driven vessel. You are in a traffic separation scheme and proceeding in an appropriate traffic lane in the general direction of traffic flow for that lane. You see these lights. What do the lights indicate to you?

**A.** It is a sailing vessel showing optional lights, starboard side on, underway.

**Q.** You see this vessel 40° on the port bow. What action will you take?

**A.** I will take a series of compass bearings of an approaching vessel and, if the compass bearing does not change appreciably, then the risk of collision exists.

**Q.** The bearing is constant and range decreasing. What action will you take now?

**A.** As per Rule 10(j), a sailing vessel shall not impede the safe passage of a power-driven vessel following a traffic lane. I would expect the sailing vessel to take action to avoid collision.

**Q.** She is not taking any action and getting closer. What would you do now?

**A.** I will sound at least five short and rapid blasts on the whistle to draw her attention. At night I will also use the Aldis lamp to draw her attention. I will also inform Vessel Traffic Services (VTS).

**Q.** She is still not taking any action and getting even closer. What will be your action now?

**A.** I will sound one short blast and then make a broad alteration of the course to starboard and pass well clear of the other vessel and, if necessary, go into the inshore traffic zone to avoid collision.

**Q.** Who is give-way in this situation?

**A.** As per Rule 18, the power-driven vessel is the give-way vessel.

**Q.** Would you take the same action if you were crossing the lane and the sailing vessel was following the lane?

**A.** Rule 10(j) only applies when a power-driven vessel is following a lane and a sailing vessel is impeding her passage. In the case of a power-driven vessel crossing the lane then Rule 18 applies here.

## SITUATION 05

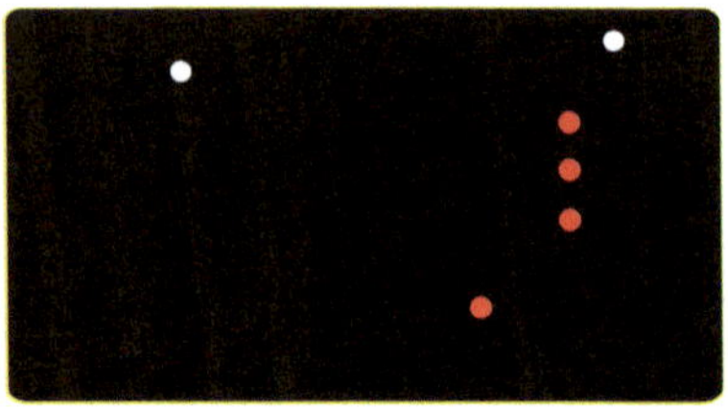

**Q.** You are the OOW of a power-driven vessel. You are in a traffic separation scheme and proceeding in an appropriate traffic lane in the general direction of traffic flow for that lane. You see these lights. What do the lights indicate to you?

**A.** A vessel constrained by her draught, probably 50m or more in length, port side on, underway.

**Q.** You see this vessel four points on your starboard bow. What action will you take?

**A.** I will take a series of compass bearings of an approaching vessel and, if the compass bearing does not change appreciably, then the risk of collision exists.

**Q.** The bearing is constant and range decreasing. What action will you take now?

**A.** As per Rule 18, I shall, if the circumstances of the case admit, avoid impeding the safe passage of a vessel constrained by her draught. In this case I will look around and reduce my speed and let her pass ahead of me.

**Q.** But there is a vessel behind you. You can't reduce the speed. What will be your action now?

**A.** I will look around and sound one short blast and then make a broad alteration of the course to starboard and pass well clear of the other vessel and go into the inshore traffic zone to avoid collision.

**Q.** But you are in traffic separation scheme. Can you go into the inshore traffic zone?

**A.** Rule 10 does not relieve any vessel of their obligations under any other rule. To avoid collision I shall alter course to starboard and, if necessary, go into the inshore traffic zone.

## SITUATION 06

**Q.** You are the OOW of a power-driven vessel. You are in a traffic separation scheme and proceeding in an appropriate traffic lane in the general direction of traffic flow for that lane. You see these lights. What do the lights indicate to you?

**A.** A power-driven vessel constrained by her draught, probably 50m or more in length, seen from the starboard side, underway.

**Q.** You see this vessel four points on the port bow. What action will you take?

**A.** I will take a series of compass bearings of an approaching vessel and, if the compass bearing does not change appreciably, then the risk of collision exists.

**Q.** The bearing is constant and range decreasing. What action will you take now?

**A.** As per Rule 18, I shall, if the circumstances of case admit, avoid impeding the safe passage of a vessel constrained by her draught. In this case I will look around and reduce my speed and let her pass ahead of me.

**Q.** Who is give-way in this situation?

**A.** The power-driven vessel constrained by her draught is give-way in this situation.

## SITUATION 07

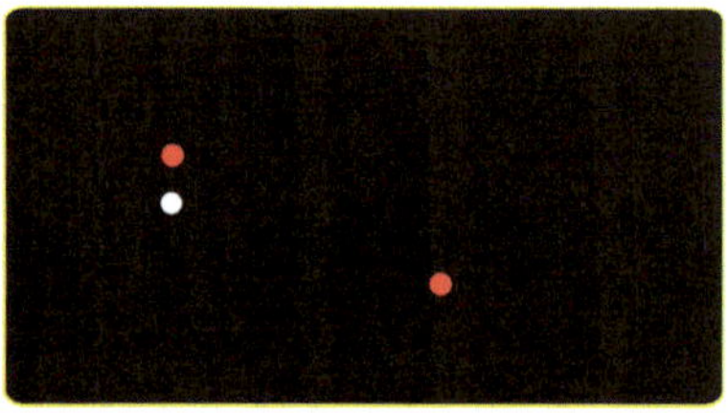

**Q.** You are the OOW of a power-driven vessel. You are in the traffic separation scheme and proceeding in an appropriate traffic lane in the general direction of traffic flow for that lane. You see these lights. What do the lights indicate to you?

**A.** A vessel engaged in fishing, other than trawling, port side on, making way through the water.

**Q.** You see this vessel 30° on the starboard bow. What action will you take?

**A.** I will take a series of compass bearings of an approaching vessel and, if the compass bearing does not change appreciably, then the risk of collision exists.

**Q.** The bearing is constant and range decreasing. What action will you take now?

**A.** As per Rule 10(i), a vessel engaged in fishing shall not impede the passage of any vessel following a traffic lane, so I would expect her to take action.

**Q.** She is not taking any action. What will you do now?
**Q.** I will sound at least five short and rapid blasts to draw her attention.
**Q.** She is still not taking any action and getting even closer. What action will you take now?
**A.** I will sound one short blast and then make a broad alteration of the course to starboard side and pass well clear of the other vessel and, if necessary, go into the inshore traffic zone to avoid collision.
**Q.** Who is give-way in this situation?
**A.** As per Rule 18, the power-driven vessel is the give-way vessel.

## SITUATION 08

**Q.** On another day, you are the OOW of a power-driven vessel constrained by her draught in a TSS following a traffic lane. You see a sailing vessel four points on the port bow. What action will you take?
**A.** I will take a series of compass bearings of an approaching vessel and, if the compass bearing does not change appreciably, then the risk of collision exists.
**Q.** The bearing is constant and range decreasing. What action will you take now?
**A.** As per Rule 10(j), a sailing vessel shall not impede the safe passage of a power-driven vessel following a traffic lane. So, I would expect the sailing vessel to take action to avoid collision.

## SITUATION 09

**Q.** On another day, you are the OOW of a power-driven vessel constrained by her draught crossing the traffic lane of a TSS. You see a sailing vessel four points on the starboard bow. What action will you take?

**A.** I will take a series of compass bearings of an approaching vessel and, if the compass bearing does not change appreciably, then the risk of collision exists.

**Q.** The bearing is constant and range decreasing. What action will you take now?

**A.** As per Rule 18(d), any vessel other than a vessel not under command or a vessel restricted in her ability to manoeuvre shall, if the circumstances of the case admit, avoid impeding the safe passage of a vessel constrained by her draught. So, I would expect the sailing vessel to take action to avoid collision.

## SITUATION 10

**Q.** You are the OOW of a power-driven vessel. You are in the traffic separation scheme and proceeding in an appropriate traffic lane in the general direction of traffic flow for that lane. You see these lights. What do the lights indicate to you?

**A.** A vessel engaged in fishing, other than trawling, seen from astern, making way through the water.

**Q.** You see this vessel right ahead. What action will you take?

**A.** I will take a series of compass bearings of an approaching vessel and, if the compass bearing does not change appreciably, then the risk of collision exists.

**Q.** The bearing is constant and range decreasing. What action will you take now?

**A.** As per Rule 13, my vessel is overtaking a vessel engaged in fishing, so it's my duty to keep clear.

**Q.** But it's the vessel engaged in fishing that shall not impede the passage of any vessel following the traffic lane. Would you not expect her to take action?

**A.** As per Rule 13, notwithstanding anything contained in the rules of Part B, Sections I and II, any vessel overtaking any other shall keep out of the way of the vessel being overtaken.

## SITUATION 11

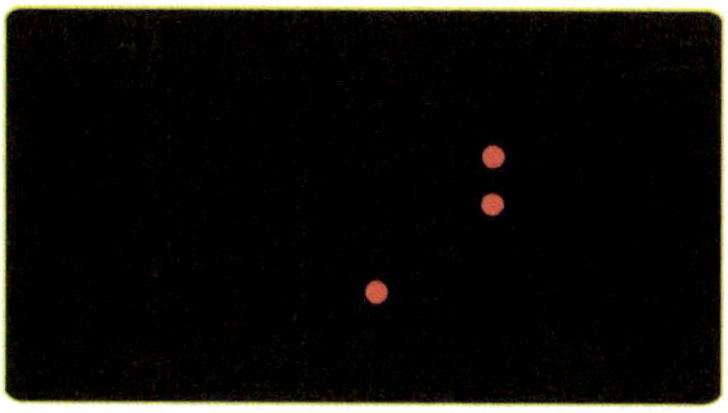

**Q.** You are the OOW of a power-driven vessel in a narrow channel. You see these lights. What do the lights indicate to you?

**A.** A vessel not under command, port side on, making way through the channel.

**Q.** You see this vessel 45° on the starboard bow. What action will you take?

**A.** I will take a series of compass bearings of an approaching vessel and, if the compass bearing does not change appreciably, then the risk of collision exists.

**Q.** The bearing is constant and range decreasing. What action will you take now?

**A.** As per Rule 9(d), a vessel shall not cross a narrow channel or fairway if such crossing impedes the passage of a vessel which can safely navigate only within such a channel or fairway. So, I would expect the vessel not under command to take action to avoid collision.

**Q.** What if she does not take action?

**A.** As per Rule 9(d), I will sound at least five or more short and rapid blasts to draw her attention.

## SITUATION 12

**Q.** Which side of a narrow channel will you keep your vessel?

**A.** I will keep to the starboard side.

**Q.** Which types of vessel shall not impede your passage when you can only navigate safely within a narrow channel?

**A.** As per Rule 9(b and c), a sailing vessel, a vessel less than 20m in length and a vessel engaged in fishing shall not impede the safe passage of a vessel navigating within a narrow channel.

**Q.** A sailing vessel is navigating within a narrow channel. Can a vessel engaged in fishing impede her passage?

**A.** As per Rule 9(c), a vessel engaged in fishing shall not impede the passage of any other vessel navigating within a narrow channel or fairway.

**Q.** You are OOW of a power-driven vessel and navigating within a narrow channel. There is a power-driven vessel on your port side crossing the channel. What action will you take?

**A.** As per Rule 9(d), a vessel shall not cross the narrow channel if such crossing impedes the passage of a vessel which can safely navigate within the narrow channel. I would expect the other vessel to take action.

**Q.** What if she does not take action?

**A.** In this case, as per Rule 9(d), I will sound at least five short and rapid blasts.

**Q.** How would you overtake a vessel within a narrow channel?

**A.** I would seek permission of the vessel being overtaken by sounding an appropriate signal and wait for the response.

**Q.** What is the appropriate signal?

**A.** Two prolonged blasted blasts followed by one short blast to mean 'I intend to overtake you on your starboard side'; or two prolonged blasted blasts followed by two short blasts to mean 'I intend to overtake you on your port side.'

# Restricted visibility situations

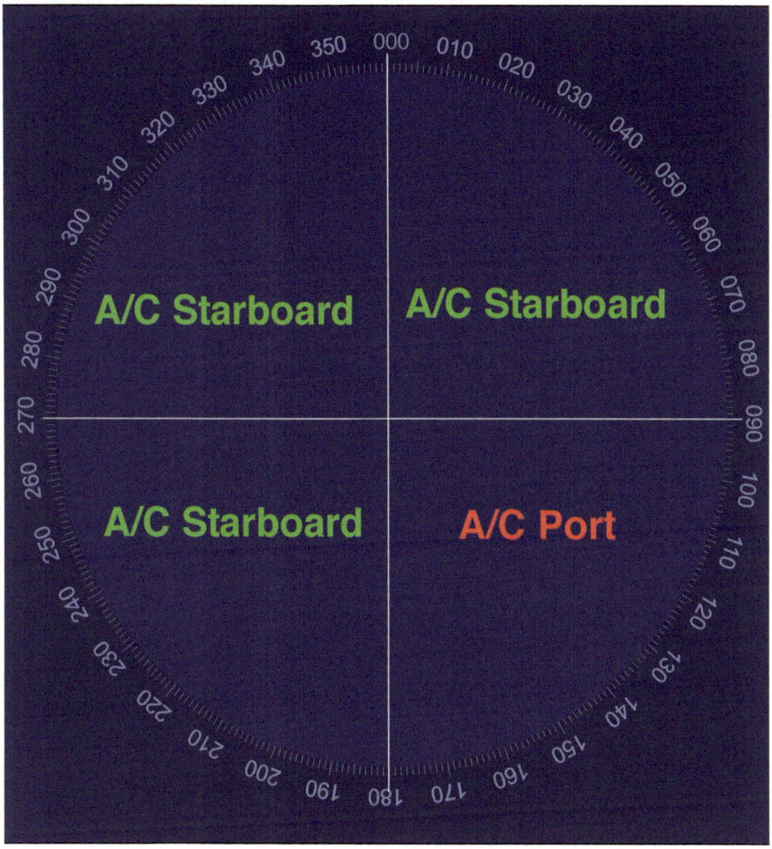


*A Practical Guide to the Rules of the Road, Third edition.* 978-1-138-84389-9.

## SITUATION 01

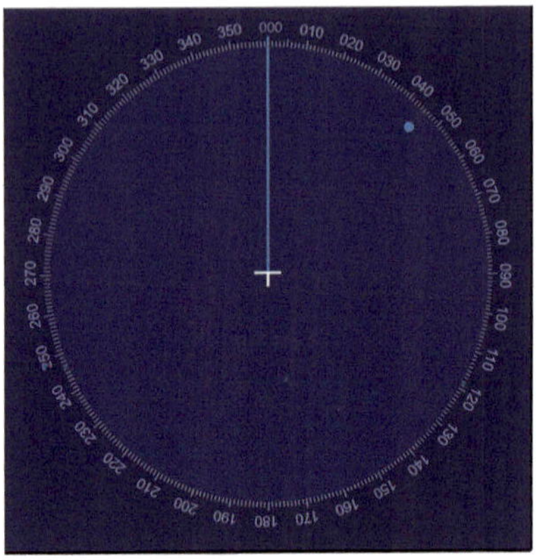

**Q.** You are the OOW of a power-driven vessel. The vessel is proceeding in restricted visibility. You observe one radar target at four points on the starboard bow. What action will you take?

**A.** I will start systematic plot to ascertain if risk of collision exists.

**Q.** How many observations will you take?

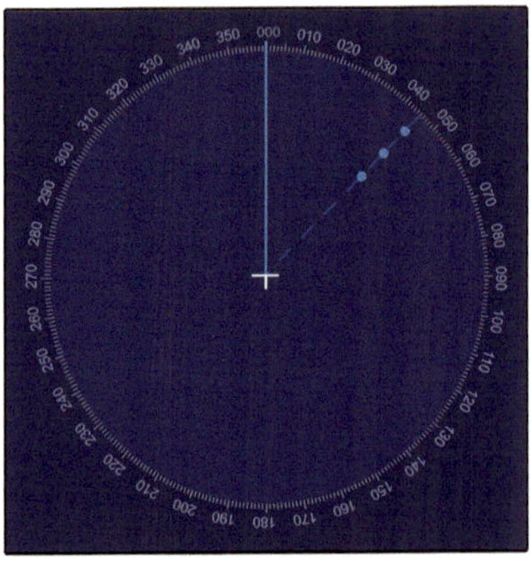

**A.** I will take at least three equal-time-interval observations.

**Q.** After three observations the plot shows bearings constant and range decreasing. What action will you take now?

**A.** As per Rule 19 a vessel shall avoid altering course to port for a vessel forward of the beam. In this case I will make a broad alteration of course to starboard and start a new plot.

**Q.** Will you sound any manoeuvring signal while altering?

**A.** No. The vessel will only sound restricted visibility signals. I am already sounding one prolonged blast at intervals of not more than two minutes.

## SITUATION 02

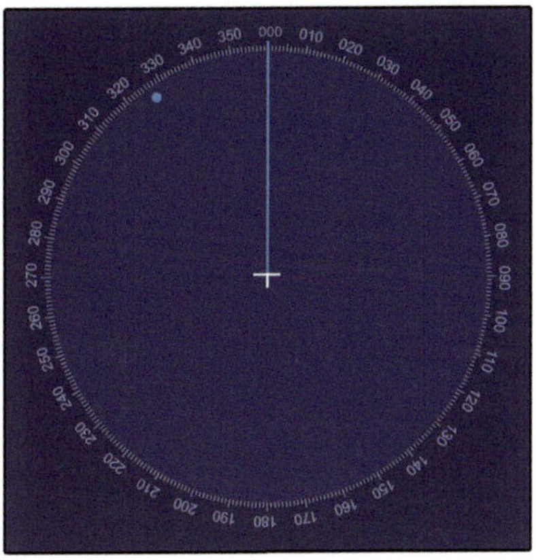


**Q.** You are the OOW of a power-driven vessel. The vessel is proceeding in restricted visibility. You observe one radar target at three points on the port bow. What action will you take?

**A.** I will start a systematic plot to ascertain if risk of collision exists.

**Q.** How many observations will you take?

**A.** I will take at least three equal-time-interval observations.

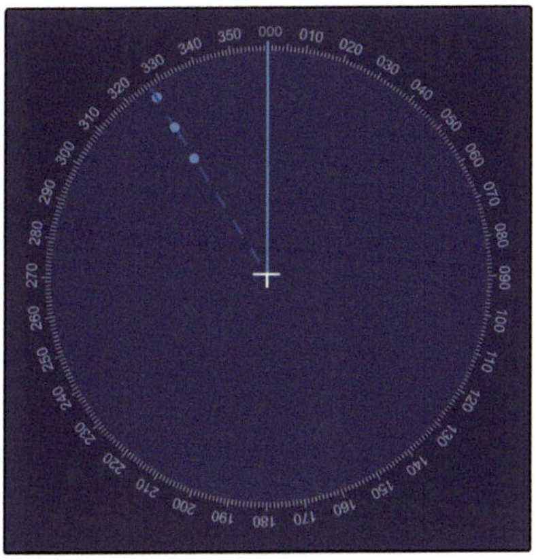


**Q.** Why do you need three equal-time-interval observations?

**A.** Three linear observations with equal distances between them shows that the target is on the same course and speed during the period of observations.

**Q.** After three observations the plot shows bearings constant and range decreasing. What action will you take now?

**A.** As per Rule 19 a vessel shall avoid altering course to port for a vessel forward of the beam. In this case I will make a broad alteration of course to starboard and start a new plot.

## SITUATION 03

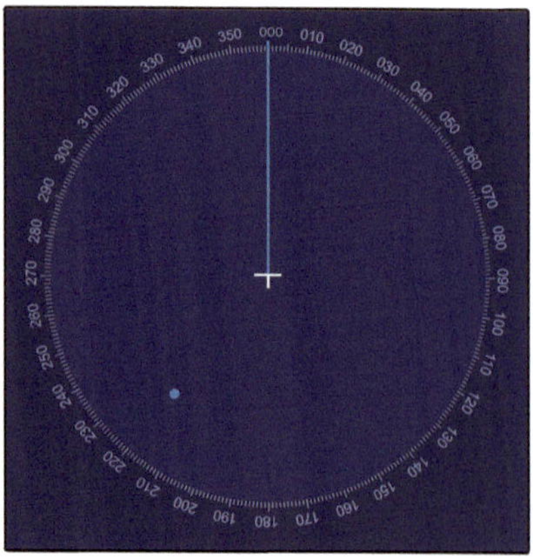

**Q.** You are the OOW of a power-driven vessel. The vessel is proceeding in restricted visibility. You observe one target on the port quarter. What action will you take?

**A.** I will start a systematic plot to ascertain if risk of collision exists.

**Q.** How many observations will you take?

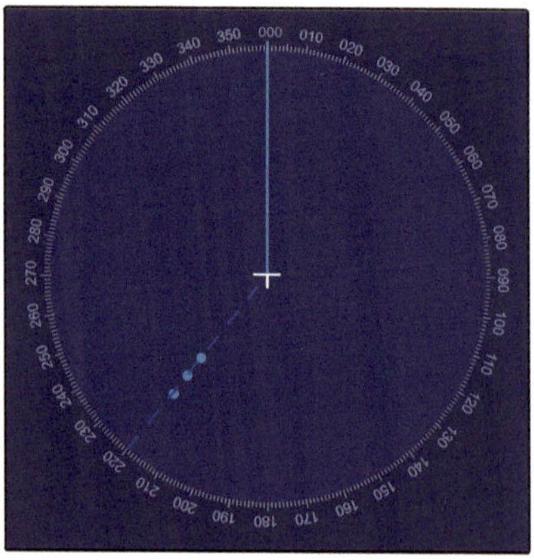

**A.** I will take at least three equal-time-interval observations.

**Q.** After three observations the plot shows bearings constant and range decreasing. What action will you take now?

**A.** As per Rule 19 a vessel shall avoid altering course towards a vessel abaft the beam. In this case I will make a broad alteration of course to starboard and start a new plot.

## SITUATION 04

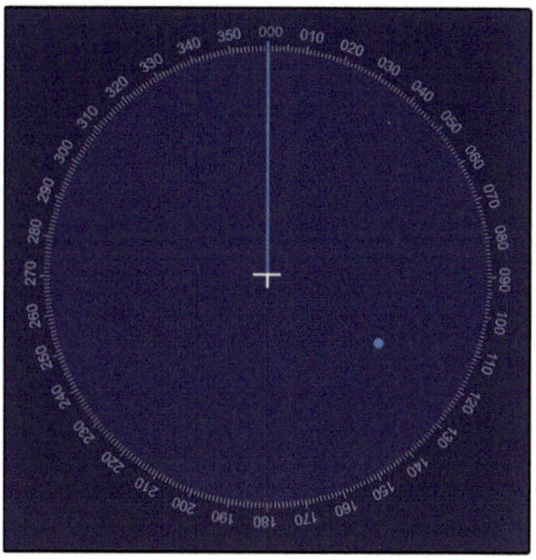

**Q.** You are the OOW of a power-driven vessel. The vessel is proceeding in restricted visibility. You observe one target on the starboard quarter. What action will you take?

**A.** I will start a systematic plot to ascertain if risk of collision exists.

**Q.** How many observations will you take?

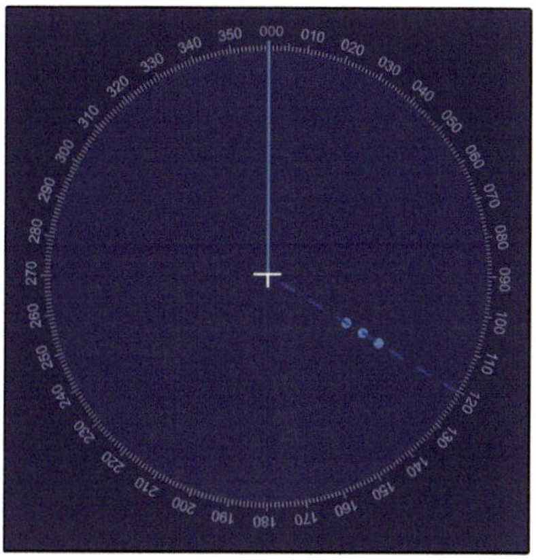

**A.** I will take at least three equal-time-interval observations.

**Q.** After three observations the plot shows bearings constant and range decreasing. What action will you take now?

**A.** As per Rule 19 a vessel shall avoid altering course towards a vessel abaft the beam. In this case I will make a broad alteration of course to port and start a new plot.

## SITUATION 05

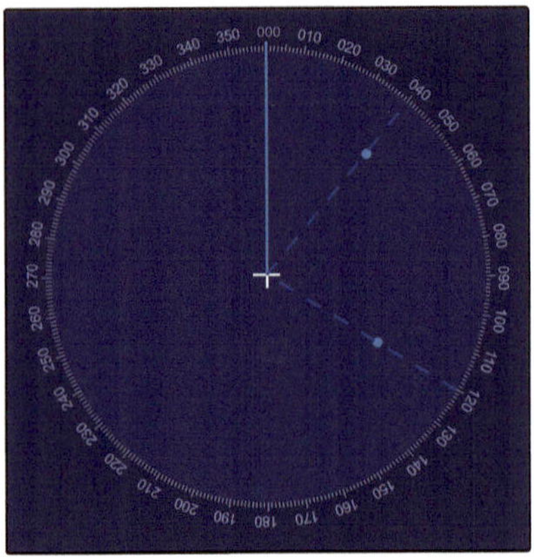


**Q.** You are the OOW of a power-driven vessel. The vessel is proceeding in restricted visibility. You observe two targets on the radar. One target is on the starboard bow and the other on the starboard quarter. Risk of collision exists with both targets. What action will you take?

**A.** As per Rule 19, a vessel shall avoid altering course to port for a vessel forward of the beam and shall also avoid altering towards the vessel for a vessel abaft the beam. In this case since I cannot alter course to port for the vessel forward of the beam and I cannot alter course to starboard for the vessel on the starboard quarter. In this situation I will stop and take all way off, and change the sound signal from one prolonged blast to two prolonged blasts and continue to monitor the situation.

## SITUATION 06

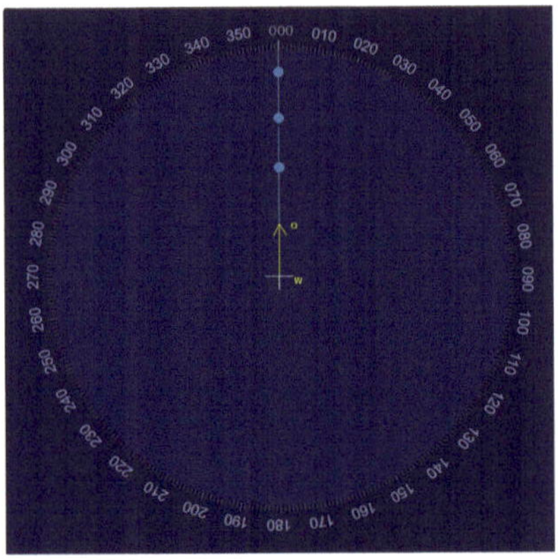


**Q.** You are the OOW of a power-driven vessel. The vessel is proceeding in restricted visibility. You observe one target right ahead. What action will you take?

**A.** I will start a systematic plot to ascertain if risk of collision exists.

**Q.** After three observations the plot shows the bearings constant and range decreasing. What action will you take now?

**A.** As per Rule 19 a vessel shall avoid altering course to port for a vessel forward of the beam. In this case I will make a broad alteration of course to starboard and start a new plot.

**Q.** Does Rule 14 apply here?

**A.** No. Rule 14 is under Section 2 of Part B which is 'Conduct of vessels in sight of one another'. With restricted visibility the rules of Section 2 of Part B do not apply.

**Q.** So which rules of part B would you apply in restricted visibility?

**A.** All rules of Section 1, 'Conduct of vessels under any condition of visibility', and Rule 19 of Section 3 apply here.

## SITUATION 07

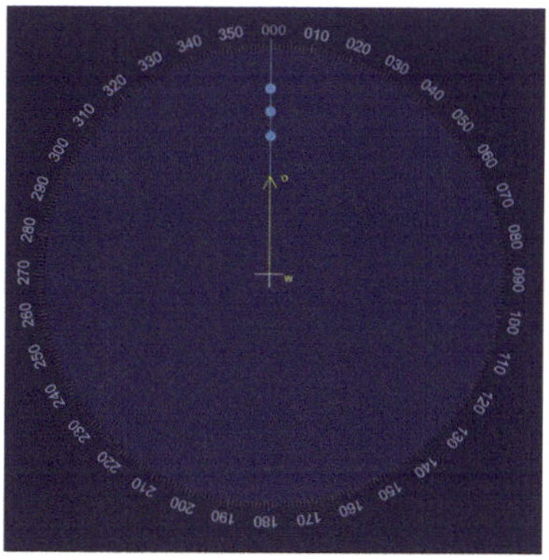


**Q.** You are the OOW of a power-driven vessel. The vessel is proceeding in restricted visibility. You observe one target right ahead. What action will you take?

**A.** I will start a systematic plot to ascertain if risk of collision exists.

**Q.** After three observations the plot shows the bearings constant and range decreasing. What action will you take now?

**A.** As per Rule 19(d) a vessel shall avoid altering course to port for a vessel forward of the beam other than vessel being overtaken. WO vector is bigger than OA vector and this is an overtaking situation. In this case I will make a broad alteration of course to starboard and start a new plot.

**Q.** Can you alter course to port?

**A.** Yes, I can alter course to port. As per Rule 19(d) I shall avoid altering course to port for a vessel forward of the beam other than vessel being overtaken and this is overtaking situation.

**Q.** Does Rule 13 apply here?

**A.** No sir, Rule 13 does not apply in restricted visibility (Section III). Rule 13 only overrides all the rules in Part B, Sections I and II.

## SITUATION 08

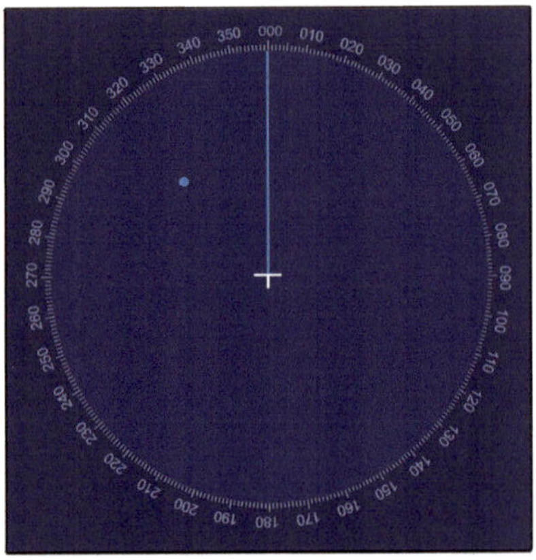

**Q.** You are the OOW of a power-driven vessel. The vessel is proceeding in restricted visibility. You observe a radar target on the port bow. What action will you take?

**A.** I will start a systematic plot to ascertain if risk of collision exists.

**Q.** How many observations will you take?

**A.** I will take at least three equal-time-interval observations.

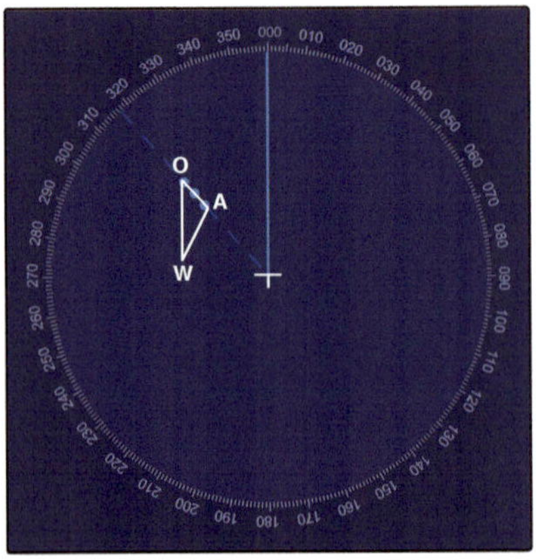


**Q.** After three observations, the plot shows bearings constant and range decreasing. Will you have a look at the radar plot triangle and take action?

**A.** Yes. The radar plot shows that I am the overtaking vessel. I will make a broad alteration of course to starboard and continue to plot the target.

**Q.** Can you alter course to port?

**A.** Yes sir. Rule 19(d)(i) states that a vessel shall avoid altering course to port for a vessel forward of the beam other than vessel being overtaken. Since I am the overtaking vessel, I can alter course to either side. But in this situation I prefer to alter course to starboard as I am on the starboard quarter of the other vessel and she will most probably alter course to port.

## SITUATION 09

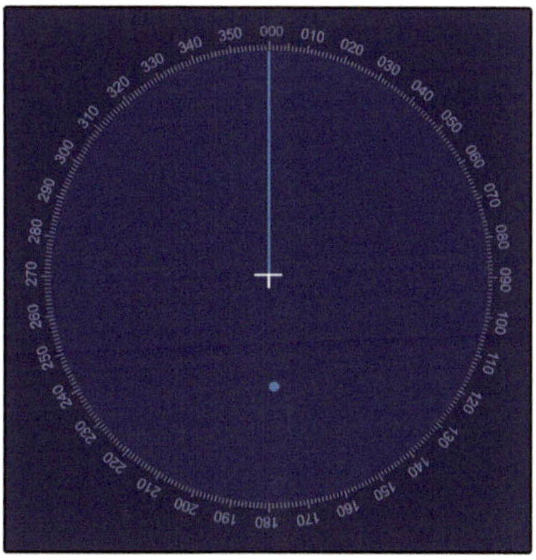


**Q.** You are the OOW of a power-driven vessel. The vessel is proceeding in restricted visibility. You observe one target right astern. What action will you take?

**A.** I will start a systematic plot to ascertain if risk of collision exists.

**Q.** After three observations, the plot shows bearings constant and range decreasing. What action will you take now?

**A.** As per Rule 19, a vessel shall avoid altering course towards the vessel for a vessel abaft the beam. In this case since vessel is neither side of my vessel, but right astern, I can therefore alter course to any side. The other vessel is overtaking, and I would expect her to alter course to starboard, so that she can keep her starboard side clear. Therefore I will make a broad alteration of course to port and start a new plot.

**Q.** Can she alter course to port?

**A.** Yes she can. Rule 19(d)(i) states that a vessel shall avoid altering course to port for a vessel forward of the beam, other than for a vessel being overtaken. The other vessel is overtaking me.

## SITUATION 10

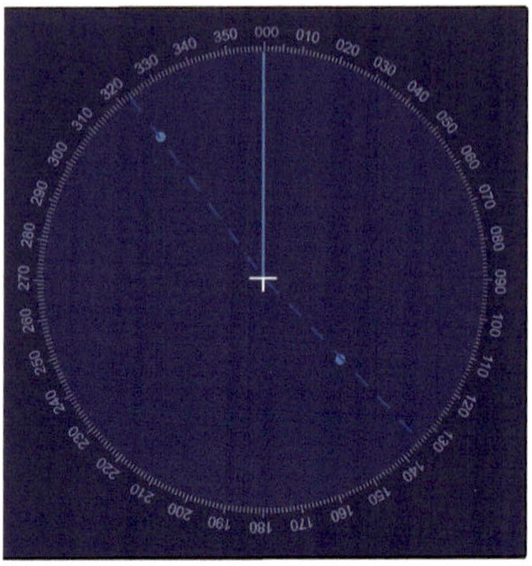

**Q.** You are the OOW of a power-driven vessel. The vessel is proceeding in restricted visibility. You observe two targets on the radar. One target is on the port bow and the other on the starboard quarter. What action will you take?

**A.** I will start a systematic plot to ascertain if a risk of collision exists.

**Q.** How many observations will you take?

**A.** I will take at least three equal-time-interval observations.

**Q.** After three observations, the plot is like this picture. What action will you take now?

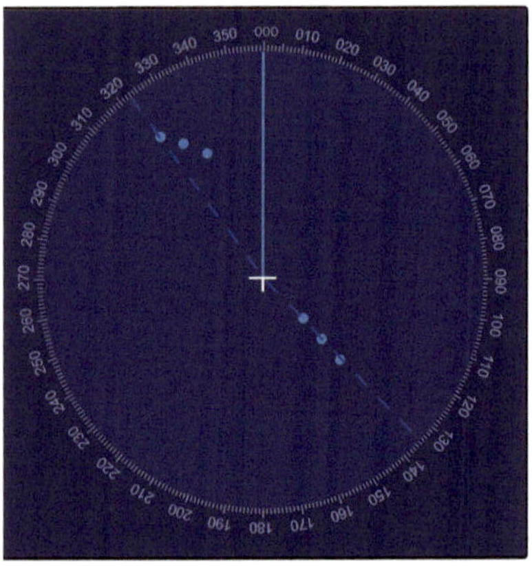

**A.** I would like to quote rule 19(d) here:

*(d)* *A vessel which detects by radar alone the presence of another vessel shall determine* if a close-quarters situation is developing and/or risk of collision

exists. If so, *she shall take avoiding action in ample time, provided that when such action consists of an alteration of course, so far as possible the following shall be avoided:*

*(i) an alteration of course to port for a vessel forward of the beam, other than for a vessel being overtaken;*

*(ii) an alteration of course towards a vessel abeam or abaft the beam.*

Since there is no risk of collision with the target on the port bow, I will take action for the target on the starboard quarter. I shall avoid altering course towards the vessel abaft the beam. In this case I will make a broad alteration of course to port and continue to plot both targets and monitor the situation.

## SITUATION 11

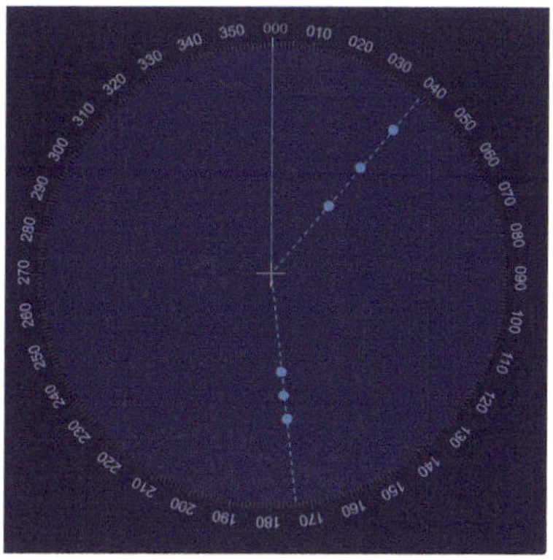

**Q.** You are the OOW of a power-driven vessel. The vessel is proceeding in restricted visibility. You observe two targets on the radar. One target is on the starboard bow and the other on the starboard quarter. What action will you take?

**A.** I will start a systematic plot to ascertain if a risk of collision exists.

**Q.** After three observations, the plot is like this picture. What action will you take now?

**A.** In this case, as per Rule 19(d)(i) I shall avoid altering course to port for the vessel forward of the beam which is the target on the starboard bow; and as per Rule 19(d)(ii) I shall avoid altering course towards the vessel abaft the beam, hence I shall avoid altering course to starboard for the vessel on my starboard quarter. The target on the starboard quarter is very fine on the astern and so reducing speed will not solve the problem. I can see that the TCPA of the target on the starboard bow is much earlier than the TCPA of the target on the starboard quarter, so I will first deal with the target on the starboard bow and make a broad alteration of course to starboard and re-plot both targets.

**Q.** Rule 19 does not state that you have to consider TCPA of the targets when selecting your action.

**A.** This action has been taken with due regard to observance of good seamanship (Rule 8(a)).

# Annexes

## ANNEX I: POSITIONING AND TECHNICAL DETAILS OF LIGHTS AND SHAPES

### 1 Definition

The term 'height above the hull' means height above the uppermost continuous deck. This height shall be measured from the position vertically beneath the location of the light.

### 2 Vertical positioning and spacing of lights

(a) On a power-driven vessel of 20 metres or more in length the masthead lights shall be placed as follows:
   (i) the forward masthead light, or if only one masthead light is carried, then that light, at a height above the hull of not less than 6 metres, and, if the breadth of the vessel exceeds 6 metres, then at a height above the hull not less than such breadth, so however that the light need not be placed at a greater height above the hull than 12 metres;
   (ii) when two masthead lights are carried the after one shall be at least 4.5 metres vertically higher than the forward one.

(b) The vertical separation of masthead lights of power-driven vessels shall be such that in all normal conditions of trim the after light will be seen over and separate from the forward light at a distance of 1,000 metres from the stem when viewed from sea-level.

(c) The masthead light of a power-driven vessel of 12 metres but less than 20 metres in length shall be placed at a height above the gunwale of not less than 2.5 metres.

(d) A power-driven vessel of less than 12 metres in length may carry the uppermost light at a height of less than 2.5 metres above the gunwale. When however a masthead

*A Practical Guide to the Rules of the Road, Third edition.* 978-1-138-84389-9.

light is carried in addition to sidelights and a sternlight or the all-round light prescribed in Rule 23(d)(i) is carried in addition to sidelights, then such masthead light or all-round light shall be carried at least 1 metre higher than the sidelights.

(e) One of the two or three masthead lights prescribed for a power-driven vessel when engaged in towing or pushing another vessel shall be placed in the same position as either the forward masthead light or the after masthead light; provided that, if carried on the aftermast, the lowest after masthead light shall be at least 4.5 metres vertically higher than the forward masthead light.

(f) (i) The masthead light or lights prescribed in Rule 23(a) shall be so placed as to be above and clear of all other lights and obstructions except as described in sub-paragraph (ii).

(ii) When it is impracticable to carry the all-round lights prescribed by Rule 27(b)(i) or Rule 28 below the masthead lights, they may be carried above the after masthead light(s) or vertically in between the forward masthead light(s) and the after masthead light(s) provided that in the latter case the requirement of Section 3(c) of this Annex shall be complied with.

(g) The sidelights of a power-driven vessel shall be placed at a height above the hull not greater than three-quarters of that of the forward masthead light. They shall not be so low as to be interfered with by deck lights.

(h) The sidelights, if in a combined lantern and carried on a power-driven vessel of less than 20 metres in length, shall be placed not less than 1 metre below the masthead light.

(i) When the rules prescribe two or three lights to be carried in a vertical line, they shall be spaced as follows:

(i) on a vessel of 20 metres in length or more such lights shall be spaced not less than 2 metres apart, and the lowest of these lights shall, except where a towing light is required, be placed at a height of not less than 4 metres above the hull;

(ii) on a vessel of less than 20 metres in length such lights shall be spaced not less than 1 metre apart and the lowest of these lights shall, except where a towing light is required, be placed at a height of not less than 2 metres above the gunwale;

(iii) when three lights are carried they shall be equally spaced.

(j) The lower of the two all-round lights prescribed for a vessel when engaged in fishing shall be at a height above the sidelights not less than twice the distance between the two vertical lights.

(k) The forward anchor light prescribed in Rule 30(a)(i), when two are carried, shall not be less than 4.5 metres above the after one. On a vessel of 50 metres or more in length this forward anchor light shall be placed at a height of not less than 6 metres above the hull.

## 3 Horizontal positioning and spacing of lights

(a) When two masthead lights are prescribed for a power-driven vessel, the horizontal distance between them shall not be less than one-half of the length of the vessel but need not be more than 100 metres. The forward light shall be placed not more than one-quarter of the length of the vessel from the stem.

(b) On a power-driven vessel of 20 metres or more in length the sidelights shall not be placed in front of the forward masthead lights. They shall be placed at or near the side of the vessel.

(c) When the lights prescribed in Rule 27(b)(i) or Rule 28 are placed vertically between the forward masthead light(s) and the after masthead light(s) these all-round lights shall be placed at a horizontal distance of not less than 2 metres from the fore and aft centreline of the vessels in the athwartship direction.

(d) When only one masthead light is prescribed for a power-driven vessel, this light shall be exhibited forward of amidships; except that a vessel of less than 20 metres in length need not exhibit this light forward of amidships but shall exhibit it as far forward as is practicable.

## 4 Details of location of direction-indicating lights for fishing vessels, dredgers and vessels engaged in underwater operations

(a) The light indicating the direction of the outlying gear from a vessel engaged in fishing as prescribed in Rule 26(c)(ii) shall be placed at a horizontal distance of not less than 2 metres and not more than 6 metres away from the two all-round red and white lights. This light shall be placed not higher than the all-round white light prescribed in Rule 26(c)(i) and not lower than the sidelights.

(b) The lights and shapes on a vessel engaged in dredging or underwater operations to indicate the obstructed side and/or the side on which it is safe to pass, as prescribed in Rule 27(d)(i) and (ii), shall be placed at the maximum practical horizontal distance, but in no case less than 2 metres, from the lights or shapes prescribed in Rule 27(b)(i) and (ii). In no case shall the upper of these lights or shapes be at a greater height than the lower of the three lights or shapes prescribed in Rule 27(b)(i) and (ii).

## 5 Screens for sidelights

The sidelights of vessels of 20 metres or more in length shall be fitted with inboard screens painted matt black, and meeting the requirements of Section 9 of this Annex. On vessels of less than 20 metres in length the sidelights, if necessary to meet the requirements of Section 9 of this Annex, shall be fitted with inboard matt black screens. With a combined lantern, using a single vertical filament and a very narrow division between the green and red sections, external screens need not be fitted.

## 6 Shapes

(a) Shapes shall be black and of the following sizes:

(i) a ball shall have a diameter of not less than 0.6 metre;
(ii) a cone shall have a base diameter of not less than 0.6 metre and a height equal to its diameter;
(iii) a cylinder shall have a diameter of at least 0.6 metre and a height of twice its diameter
(iv) a diamond shape shall consist of two cones as defined in (ii) above having a common base.

(b) The vertical distance between shapes shall be at least 1.5 metres.

(c) In a vessel of less than 20 metres in length shapes of lesser dimensions but commensurate with the size of the vessel may be used and the distance apart may be correspondingly reduced.

## 7 Colour specification of lights

The chromaticity of all navigation lights shall conform to the following standards, which lie within the boundaries of the area of the diagram specified for each colour by the International Commission on Illumination (CIE).

The boundaries of the area for each colour are given by indicating the corner co-ordinates, which are as follows:

*Table AX1* Colour specification of lights: corner coordinates

| | | | | | | |
|---|---|---|---|---|---|---|
| (i) *White* | | | | | | |
| x | 0.525 | 0.525 | 0.452 | 0.310 | 0.310 | 0.443 |
| y | 0.382 | 0.440 | 0.440 | 0.348 | 0.283 | 0.382 |
| (ii) *Green* | | | | | | |
| x | 0.028 | 0.009 | 0.300 | 0.203 | | |
| y | 0.385 | 0.723 | 0.511 | 0.356 | | |
| (iii) *Red* | | | | | | |
| x | 0.680 | 0.660 | 0.735 | 0.721 | | |
| y | 0.320 | 0.320 | 0.265 | 0.259 | | |
| (iv) *Yellow* | | | | | | |
| x | 0.612 | 0.618 | 0.575 | 0.575 | | |
| y | 0.382 | 0.382 | 0.425 | 0.406 | | |

## 8 Intensity of lights

(a) The minimum luminous intensity of lights shall be calculated by using

$$I = 3.43 \times 10^{6} \times T \times D^{2} \times K^{-D}$$

where I is luminous intensity in candelas under service conditions, T is threshold factor $2 \times 10^{-7}$ lux, D is range of visibility (luminous range) of the light in nautical miles, and K is atmospheric transmissivity. For prescribed lights the value of K shall be 0.8, corresponding to a meteorological visibility of approximately 13 nautical miles.

• • •

(b) A selection of figures derived from the formula is given in the following table:

*Table AX2* Figures derived from the minimum luminous intensity formula

| *Range of visibility (luminous range) of light in nautical miles D* | *Luminous intensity of light in candelas for K = 0.8* |
|---|---|
| 1 | 0.9 |
| 2 | 4.3 |
| 3 | 12 |
| 4 | 27 |
| 5 | 52 |
| 6 | 94 |

*Note*: The maximum luminous intensity of navigation lights should be limited to avoid undue glare. This shall not be achieved by a variable control of the luminous intensity.

## 9 Horizontal sectors

(a) (i) In the forward direction, sidelights as fitted on the vessel shall show the minimum required intensities. The intensities shall decrease to reach practical cut-off between 1 degree and 3 degrees outside the prescribed sectors.

(ii) For stern lights and masthead lights at 22.5 degrees abaft the beam for sidelights, the minimum required intensities shall be maintained over the arc of the horizon up to 5 degrees within the limits of the sectors prescribed in Rule 21. From 5 degrees within the prescribed sectors the intensity may decrease by 50 per cent up to the prescribed limits: it shall decrease steadily to reach practical cutoff at not more than 5 degrees outside the prescribed sectors.

(b) (i) All-round lights shall be so located as not to be obscured by masts, topmasts or structures within angular sectors of more than 6 degrees, except anchor lights prescribed in Rule 30, which need not be placed at an impracticable height above the hull.

(ii) If it is impracticable to comply with paragraph (b) (i) of this section by exhibiting only one all-round light, two all-round lights shall be used suitably positioned or screened so that they appear, as far as practicable, as one light at a distance of one mile.

## 10 Vertical sectors

(a) The vertical sectors of electric lights as fitted, with the exception of lights on sailing vessels underway, shall ensure that:

(i) at least the required minimum intensity is maintained at all angles from 5 degrees above to 5 degrees below the horizontal;

(ii) at least 60 per cent of the required minimum intensity is maintained from 7.5 degrees above to 7.5 degrees below the horizontal.

(b) In the case of sailing vessels underway the vertical sectors of electric lights as fitted shall ensure that:

(i) at least the required minimum intensity is maintained at all angles from 5 degrees above to 5 degrees below the horizontal;

(ii) at least 50 per cent of the required minimum intensity is maintained from 25 degrees above to 25 degrees below the horizontal.

(c) In the case of lights other than electric these specifications shall be met as closely as possible.

## 11 Intensity of non-electric lights

Non-electric lights shall so far as practicable comply with the minimum intensities, as specified in the table given in Section 8 of this Annex.

## 12 Manoeuvring light

Notwithstanding the provisions of paragraph 2(f) of this Annex, the manoeuvring light described in Rule 34(b) shall be placed in the same fore and aft vertical plane as the masthead light or lights and, where practicable, at a minimum height of 2 metres vertically above the forward masthead light, provided that it shall be carried not less than 2 metres vertically above or below the after masthead light. On a vessel where only one masthead light is carried the manoeuvring light, if fitted, shall be carried where it can best be seen, not less than 2 metres vertically apart from the masthead light.

## 13 High speed craft*

(a) The masthead light of high speed craft may be placed at a height related to the breadth of the craft lower than that prescribed in paragraph 2(a)(i) of this Annex, provided that the base angle of the isosceles triangles formed by the sidelights and masthead light, when seen in end elevation, is not less than 27 degrees.

(b) On high speed craft of 50m or more in length, the vertical separation between foremast and mainmast light of 4.5m required by paragraph 2(a)(ii) of this annex may be modified provided that such distance shall not be less than the value determined by the following formula:

$$y = \frac{(a + 17\,\psi)}{1000}\,C + 2$$

Where $y$ is the height of the mainmast light above the foremast light in metres, $a$ is the height of the foremast light above the water surface in service condition in metres, $C$ is the trim in service condition in degrees, and $\psi$ is the horizontal separation of masthead lights in metres.

* Refer to the International Code of Safety for High-Speed Craft, 1994 and the International Code of Safety for High-Speed Craft, 2000.

## 14 Approval

The construction of lights and shapes and the installation of lights on board the vessel shall be to the satisfaction of the appropriate authority of the state whose flag the vessel is entitled to fly.

# ANNEX II: ADDITIONAL SIGNALS FOR FISHING VESSELS FISHING IN CLOSE PROXIMITY

## 1 General

The lights mentioned herein shall, if exhibited in pursuance of Rule 26(d), be placed where they can best be seen. They shall be at least 0.9 metre apart but at a lower level than lights prescribed in Rule 26(b)(i) and (c)(i). The lights shall be visible all round the horizon at a distance of at least 1 mile but at a lesser distance than the lights prescribed by these rules for fishing vessels.

## 2 Signals for trawlers

(a) Vessels of 20 metres or more in length when engaged in trawling, whether using demersal or pelagic gear, shall exhibit:
  (i) when shooting their nets, two white lights in a vertical line;
  (ii) when hauling their nets, one white light over one red light in a vertical line;
  (iii) when the net has come fast upon an obstruction, two red lights in a vertical line.

(b) Each vessel of 20 metres or more in length engaged in pair trawling shall exhibit:
  (i) by night, a searchlight directed forward and in the direction of the other vessel of the pair;
  (ii) when shooting or hauling their nets or when the nets have come fast upon an obstruction, the lights prescribed in 2(a) above.

(c) A vessel of less than 20 metres in length engaged in trawling, whether using demersal or pelagic gear or engaged in pair trawling, may exhibit the lights prescribed in paragraphs (a) or (b) of this section, as appropriate.

## 3 Signals for purse seiners

Vessels engaged in fishing with purse seine gear may exhibit two yellow lights in a vertical line. These lights shall flash alternately every second and with equal light and occultation duration. These lights may be exhibited only when the vessel is hampered by its fishing gear.

# ANNEX III: TECHNICAL DETAILS OF SOUND SIGNAL APPLIANCES

## 1 Whistles

### *(a) Frequencies and range of audibility*

The fundamental frequency of the signal shall lie within the range 70–700Hz.

The range of audibility of the signal from a whistle shall be determined by those frequencies, which may include the fundamental and/or one or more higher frequencies, which lie within the range 180–700Hz (+/– 1%) for a vessel of 20m or more in length, or 180–2,100Hz (+/– 1%) for a vessel of less than 20m in length and which provide the sound pressure levels specified in paragraph l(c) below.

### *(b) Limits of fundamental frequencies*

To ensure a wide variety of whistle characteristics, the fundamental frequency of a whistle shall be between the following limits:

(i) 70–200Hz, for a vessel 200 metres or more in length;
(ii) 130–350Hz, for a vessel 75 metres but less than 200 metres in length;
(iii) 250–700Hz, for a vessel less than 75 metres in length.

### *(c) Sound signal intensity and range of audibility*

A whistle fitted in a vessel shall provide, in the direction of maximum intensity of the whistle and at a distance of 1 metre from it, a sound pressure level in at least one 1/3rd-octave band within the range of frequencies 180–700Hz (+/– 1%) for a vessel of 20m or more in length, or 180–2,100Hz (+/– 1%) for a vessel of less than 20m in length, of not less than the appropriate figure given in the table below.

*Table AX3* Sound signal intensity and range of audibility

| *Length of vessel in metres* | *1/3rd-octave band level at 1 metre in dB referred to $2 \times 10^{-5}$ N/m²* | *Audibility range in nautical miles* |
|---|---|---|
| 200 or more | 143 | 2 |
| 75 but less than 200 | 138 | 1.5 |
| 20 but less than 75 | 130 | 1 |
| Less than 20 | 120* | 0.5 |
| | 115† | |
| | 111‡ | |

*Notes*: * When the measured frequencies lie within the range 180–450Hz
† When the measured frequencies lie within the range 450–800Hz
‡ When the measured frequencies lie within the range 800–2100Hz

The range of audibility in the table above is for information and is approximately the range at which a whistle may be heard on its forward axis with 90 per cent probability in conditions of still air on board a vessel having average background noise level at the listening posts (taken to be 68dB in the octave band centred on 250Hz and 63dB in the octave band centred on 500Hz). In practice the range at which a whistle may be heard is extremely variable and depends critically on weather conditions; the values given can be regarded as typical but under conditions of strong wind or high ambient noise level at the listening post the range may be much reduced.

### (d) Directional properties

The sound pressure level of a directional whistle shall be not more than 4dB below the prescribed sound pressure level on the axis at any direction in the horizontal plane within +/– 45 degrees of the axis. The sound pressure level at any other direction in the horizontal plane shall be not more than 10dB below the prescribed sound pressure level on the axis, so that the range in any direction will be at least half the range on the forward axis. The sound pressure level shall be measured in that 1/3rd-octave band which determines the audibility range.

### (e) *Positioning of whistles*

When a directional whistle is to be used as the only whistle on a vessel, it shall be installed with its maximum intensity directed straight ahead.

A whistle shall be placed as high as practicable on a vessel, in order to reduce interception of the emitted sound by obstructions and also to minimize hearing damage risk to personnel. The sound pressure level of the vessel's own signal at listening posts shall not exceed 110dB (A) and so far as practicable should not exceed 100dB (A).

### (f) *Fitting of more than one whistle*

If whistles are fitted at a distance apart of more than 100 metres, it shall be so arranged that they are not sounded simultaneously.

### (g) *Combined whistle systems*

If due to the presence of obstructions the sound field of a single whistle or one of the whistles referred to in paragraph (f) above is likely to have a zone of greatly reduced signal level, it is recommended that a combined whistle system be fitted so as to overcome this reduction. For the purposes of the rules a combined whistle system is to be regarded as a single whistle. The whistles of a combined system shall be located at a distance apart of not more than 100 metres and arranged to be sounded simultaneously. The frequency of any one whistle shall differ from those of the others by at least 10Hz.

## 2 Bell or gong

### (a) *Intensity of signal*

A bell or gong, or other device having similar sound characteristics shall produce a sound pressure level of not less than 110dB at a distance of 1 metre from it.

### *(b) Construction*

Bells and gongs shall be made of corrosion-resistant material and designed to give a clear tone. The diameter of the mouth of the bell shall be not less than 300mm for vessels of 20 metres or more in length. Where practicable, a power-driven bell striker is recommended to ensure constant force but manual operation shall be possible. The mass of the striker shall be not less than 3 per cent of the mass of the bell.

## 3 Approval

The construction of sound signal appliances, their performance and their installation on board the vessel shall be to the satisfaction of the appropriate authority of the state whose flag the vessel is entitled to fly.

## ANNEX IV: DISTRESS SIGNALS

1. The following signals, used or exhibited either together or separately, indicate distress and need of assistance:
   (a) a gun or other explosive signal fired at intervals of about a minute;
   (b) a continuous sounding with any fog-signalling apparatus;
   (c) rockets or shells, throwing red stars fired one at a time at short intervals;
   (d) a signal made by any signalling method consisting of the group ... – – – ... (SOS) in the Morse Code;
   (e) a signal sent by radiotelephony consisting of the spoken word 'Mayday';
   (f) the International Code Signal of distress indicated by N.C.;
   (g) a signal consisting of a square flag having above or below it a ball or anything resembling a ball;
   (h) flames on the vessel (as from a burning tar barrel, oil barrel, etc.);

(i) a rocket parachute flare or a hand flare showing a red light;
(j) a smoke signal giving off orange-coloured smoke;
(k) slowly and repeatedly raising and lowering arms outstretched to each side;
(l) a distress alert by means of digital selective calling (DSC) transmitted on:
  (i) VHF channel 70, or
  (ii) MF/HF on the frequencies 2187.5 kHz, 8414.5 kHz, 4207.5 kHz, 6312 kHz, 12577 kHz or 16804.5 kHz;
(m) a ship-to-shore distress alert transmitted by the ship's Inmarsat or other mobile satellite service provider ship earth station;
(n) signals transmitted by emergency position-indicating radio beacons;
(o) approved signals transmitted by radiocommunication systems, including survival craft radar transponders.

2. The use or exhibition of any of the foregoing signals except for the purpose of indicating distress and need of assistance and the use of other signals which may be confused with any of the above signals is prohibited.
3. Attention is drawn to the relevant sections of the International Code of Signals, the *Merchant Ship Search and Rescue Manual* and the following signals:
(a) a piece of orange-coloured canvas with either a black square and circle *or* other appropriate symbol (for identification from the air);
(b) a dye marker.

# Index